ANTHRAX

Bioterror as Fact and Fantasy

ANTHRAX

Bioterror as Fact and Fantasy

PHILIPP SARASIN

Translated by Giselle Weiss

HARVARD UNIVERSITY PRESS

Cambridge, Massachusetts, and London, England · 2006

Printed in the United States of America

Originally published as *"Anthrax": Bioterror als Phantasma.* © Suhrkamp Verlag Frankfurt am Main 2004.

Library of Congress Cataloging-in-Publication Data
Sarasin, Philipp, 1956–
[Anthrax. English]
Anthrax : bioterror as fact and fantasy / Philipp Sarasin ; translated by Giselle Weiss.
p. cm.
ISBN-13: 978-0-674-02346-8 (alk. paper)
ISBN-10: 0-674-02346-3 (alk. paper)
1. Bioterrorism—United States. 2. Anthrax—United States. 3. Postal service—United States. 4. Victims of terrorism—United States. I. Title.

HV6433.35.S271 2006
363.325′30973—dc22

Contents

ANTHRAX

Bioterror as Fact and Fantasy

Prologue: Ground Zero

"It's like a war game. The human body is a sprawling city; the spiky ball, a virus. A virus that manages to elude our outer defenses and get right inside the body has just one objective: to make more copies of itself—to multiply." Thus begins an explanation of viruses in *Killers into Cures,* a BBC documentary film produced in 2000.[1] The film shows nasty-looking, spiky projectiles zooming in on the skyscrapered shoreline of an American city—a city just like New York. Illness is depicted as an invasion of these creepy objects swooping among the high-rise buildings, attacking the body's cells—depicted as bumper cars—and immobilizing them in order to perpetuate themselves. Next, the "body" itself is devastated. In the wake of September 11, these representations seem oddly familiar—the TV images and photographs of the gaping wound at Ground Zero in October 2001 invite vivid comparisons between the city and our vulnerable bodies.

It was three *New York Times* journalists who first used the expression "gaping wound" to refer to the destruction on Manhattan's south side. In the preface to the German edition of their book *Germs: Biological Weapons, and America's Secret War*—originally published on September 11, and in German in 2002—the three shaken authors drew a connection between the "wound" in New York City and a possible biological attack.[2] They were perpetuating a small, almost imperceptible metaphorical shift that has become a fixture of discourse since the late nineteenth century: the city is a body—beautiful, thriving, pulsating with streams of people and traffic like blood through the veins and arteries, or sprawling, degenerate, sick, a place of disease and deterioration. Given the cultural, scientific, political, and technological context of September 11, the conclusion seems obvious: this terror attack was a life-threatening injury to the body of the city. There doesn't seem to be anything metaphorical about it.

What is more real than wounds? What is truer than grievous injury? What could pose a greater danger than invisible enemies—be they microbes or terrorists—insidiously infiltrating the body to destroy it from within, like the space invaders of videogames? In Richard Preston's *The Cobra Event* (1997), a science fiction thriller about a biological attack on New York City, the city is an "organism called New York City"—breathtakingly beautiful, yet infected: "The cells were people . . . Austen's patient, for the moment, was the city of New York. A couple of cells inside the patient had winked out in a mysterious way. This might be a sign of illness in the patient, or it might be nothing."[3] Only it wasn't nothing. It was already taken for granted "in Washington" that the city could become sick, that in the future, terror might come to mean bioterror. Indeed, fearful that "the terrorist attacks on New York and Washington could also have been accompanied by a biological attack," the three *Times* journalists learned on inquiring that "minutes after two jets slammed into the World Trade Center, the National Guard was mobilized . . . A 22-member unit was ordered into Manhattan to test the air for deadly germs or chemical toxins. None were found."[4]

You would have to think of it, though. Within minutes, the fact of airplanes smashing into skyscrapers somehow led people to conclude a possible bioterror attack, which was met by calling out the National Guard (while Air Force interceptor jets remained on the ground). How weird is that? How did the image of low-flying aircraft ripping a deep, dark gash in the body of the city morph into the sort of fantasy image of penetrating microbes seen in techno-scientific videogames? Why did it take only minutes for terror to become bioterror?

And it gets even more complicated: In October 2001, the first press reports appeared to the effect that White House personnel and probably also George W. Bush himself had begun taking the potent antibiotic Cipro on the morning of September 11. These reports were confirmed by the White House in 2002.[5] Were real microbes at issue on September 11? Almost a month after the attacks this perception seemed to be corroborated when five anonymous letters laced with deadly, dried anthrax spores of weaponized quality (sized less than five micrometers) infected twenty-two people, killing five.[6] Almost as a matter of course, the anthrax spores seemed to implicate the terrorism of Al Qaeda, and the bodily wounds seemed one with those that had festered for weeks at Ground Zero.

After September 11, the notion of "bioterror," a topic already current in certain military and political milieus as well as in the products of science fiction and fantasy, became a discursive hot zone in which anthrax turned into "anthrax." Not just five anthrax letters but more likely several thousand letters had been circulating from the beginning, carrying imaginary "anthrax." In other words, letters were claimed to have surfaced here and there, and the rumors sufficed to spread fear and panic. Anthrax became a metaphor. You might think that microbes would be less dangerous once they become metaphors. Anthrax may be deadly, but the word "anthrax" would simply be a figure of speech, sound without substance. This assumption is misleading, however, and this essay will show just how harmful and highly infectious metaphors such as "anthrax" can be.

Within language, metaphors act like contaminating material that transfers meaning from one realm of reality to another. This is not a particularly surprising phenomenon; rather, it is intrinsic to language itself. Metaphors are "media of exchange" among various discourses.[7] They are interfaces that structure perception in a complex way, because with them, as Lacan once stated in his typical laconic way, "It is possible to say something by saying something else." Therein lies the true power of language.[8] Even the BBC film purports to say one thing yet says another. In both instances—talking about infectious diseases and speaking of the city as a body threatened by intruders—it is specific. These images appear to be telling us the truth.

For that reason, this discursive hot zone, this semantic infection that has contaminated political discourse worldwide since September 11, deserves to be taken seriously. *The Cobra Event,* a book that was strongly recommended to then President Bill Clinton by geneticist Craig Venter and about which more will be said in this essay, also malignantly mixes "reality" and "fiction." The two are inextricably linked again and again in the expression "gaping wound" and in the analysis of events leading up to September 11. In the discursive domain of bioterror, rhetoric that invokes a "deadly hazard" stemming from a few "terror cells" in the body of our society and the need to prevent the "dispersion of biological weapons" can transform politics into political pest control. The notion of "infection," whose metaphorical link to New York was not limited to the BBC film, has become the master metaphor of western invasion scenarios. Bioterror has become a political obsession—I will call it, more precisely, a phantasm—fed by "anthrax."

The fear of letters laced with anthrax has already been forgotten, a passing hysteria. And still these letters—the real and the fictitious, those containing anthrax and those that never existed—have enormous political significance. Before the war in Iraq, the Bush administration grounded its belligerent Middle East policy predominately on the threat that so-called rogue states might provide terrorists with weapons of mass destruction. And here the administration meant neither carpet cutters nor commercial aircraft but rather nuclear, biological, and chemical weapons—weaponized anthrax spores, for instance.

I will argue that "anthrax" was crucial in extending Bush's long and unconventional "war on terror," as he pronounced it on September 20, 2001, beyond Afghanistan to Iraq. I will attempt to clarify the role played by the anthrax letters in unleashing the war on terror. Moreover, I will suggest that the mailings were possibly an inside job that helped the Bush administration to link the terror attacks of September 11 with widespread fear of weapons of mass destruction and gain support for a war against Saddam Hussein. Like any letter, the anthrax letters of autumn 2001 contained a message. It is the goal of this book to decipher that message.

To understand messages, we need to be prepared for them in some way. In that sense, "anthrax" is connected to September 11 because for years leading up to the event, countless antiterror exercises in the United States conceived terror primarily as "bioterror." The United States was expecting the anthrax letters, and the subsequent fake "anthrax" letters and "anthrax" alarms are part of a culture that dreams of bioterror. The anthrax/"anthrax" letters have a complex perceptual history that is bound up with their political function and at the same time points to their current context.

The longstanding fear of those who want to poison "us" is what primarily gave legs to the anthrax letter–related panic. Examining this history will show that the infectious poison, that is, the microbes invading our bodies, keep surreptitiously metamorphosing into the foreign bodies believed to carry the microbes. It is no accident that, following September 11, anthrax spores, themselves foreign bodies (immunologists call them antigens) raised questions about foreign bodies infiltrating the city. These bodies, which appear to be infectious as such, infect "us" simply because they are different.

Seen in this way, the events of September 11 thrust one of the central phantasms of the modern age itself into the foreground, even if just temporarily: the phantasm of the enemy as a microbe, a parasite, a partially or completely invisible pest that must be exterminated. It is possible to show, however, that this perceptual pattern reflects a fundamental error, as Slavoj Žižek argued shortly after September 11: "Whenever we encounter such a purely evil Outside, we should gather the courage to endorse the Hegelian lesson: in this pure Outside, we should recognize the distilled version of our own essence."[9] I will come back to this point.

A phantasm is a strong and very basic perceptual pattern, a sort of *idée fixe* that organizes our world view. It is thus also a shield that protects us from the real.[10] The question is: What could the phantasm of foreign bodies and of bioterror that began to circulate on September 11 possibly be protecting us from? Let me put the question another, more speculative way: What is hiding "behind" this phantasm? Political motives aside, what caused the fear of anthrax-contaminated letters to take on so incredible and pervasive a life of its own? Suppose—just suppose—that the fear-mongering political din masked a perverse desire for "anthrax." Is it possible that we hanker after the pleasure of infection, the desire to play with foreign bodies that was and is simultaneously more fantastic and more real, even more "realistic," than any fear of bioterror? Could infection be the central metaphor of globalization, and bioterror the corresponding game?

I will argue that "infection" is indeed the master metaphor for globalization, and that it enables us to decipher the disposition (or "dispositives," to use Michel Foucault's term) of threat and defense. But that implies also considering the dispositives of power which arise in this context. Foucault examined and re-examined power and its effectiveness, and thereby identified how public and other authorities handled infectious diseases. As we will see, he distinguished a leprosy model of power from a plague model, and eventually also added a smallpox model of power. The latter, developed in 1978, serves up a liberal message that is worth our attention. For it is precisely the liberality that Foucault proposed that is on the defensive today, under threat from the global war on terror.

September 11 is history, and the anthrax letters are forgotten. In Iraq, bombs and tanks long ago gave way to bloody, low-intensity warfare. And Operation Red Dawn—named for director John Milius's guerilla war movie—has landed Saddam Hussein in American detention at an undisclosed location.[11] Is it time for the historians to take over? Well, not quite: the archives are immature, and the "war on terror" is still dictating the world political agenda. This essay is thus not a historical account but rather an attempt to consider the political situation from a cultural studies perspective, in the context of history. "Cultural studies" refers to what may be for some the distasteful assumption that images and fiction, phantasms and dreams shape reality in such a way that reality and fiction become inextricably mixed. "Bioterror" is the dream dreamed by postmodern society in the throes of a self-determined state of war, and "anthrax" its wish fulfillment.

This book relies heavily on online versions of American and European newspapers, Internet websites for the White House, government agencies, research institutes, and political groups of all stripes. To some extent the verifiability of my arguments is limited by the expiration date of the sources (apart from the newspapers, which can be checked against the print versions), and eventually another sort of archival effort will be necessary.

I

On the Way to Baghdad

1

Videogames, 9/11, and the Anthrax Letters

The BBC documentary film mentioned in the Prologue conceives the body as a city—and, less overtly, the city as a body. The film compares the pathogenic microbes that threaten our lives to a flying bomb. The bomb is a spiky ball object that destroys the "cells" of the city—cars and people—as it infects them. Popular science tries to make complicated science understandable. It relates scientific phenomena to the world that we know by giving a recognizable shape to the microbiological processes taking place in our bodies. Accordingly, the BBC film portrays the microbiological entity as a visually comprehensible enemy invader entering the body like an airplane enters a city and multiplying "explosively." Up to September 11, these images were fairly intelligible; afterward, they became confusing. They seemed to suggest that the two planes struck the World Trade Center towers because the killer T cells did not shoot them early enough, or because the white blood cells (known since the time of Elie Metchnikoff, the Franco-Russian discoverer of immunology, as "soldiers of the body") failed to overwhelm the hijackers.[1]

Can one really say that, though? It could be argued that the images in the movie have nothing to do with September 11, that they are only metaphors—and forgettable ones at that—convenient analogies devised by a few science writers. But I think not. Indeed, I prefer to see this film as an example of a very eloquent, even iconic metaphor, an image that goes to the heart of a fundamental, contemporary pattern of perception. In this sense, I will use the movie image of the airborne "spiky invader" in the virtual city—the city that is New York—as a key to unraveling, at least in part, the tangled story of September 11 and the terror of the anthrax letters. Remember: airplanes are not infectious, but the hunt for pathogenic microorganisms was the first and only defensive measure taken by the American military on September 11. This logic begs to be followed, even if it leads at times through the seemingly fathomless and disembodied world of the imaginary.

Let us start with the technical side of things: the form of the representation. The BBC documentary film couches its microbiological message in the visual language of video, that is to say, computer games. The film alludes to the bulky, flying-microbe-like aliens from the first *Space Invaders* game, launched by the Japanese firm Taito in 1978. *Pac-Man,* created by Namco a year later, also evoked voracious cells, cells such as phagocytes.[2] The concept suggested feeding, or "phagocytosis," the devouring of dead cell parts or harmful bacteria in the circulation and tissue by macrophages (a kind of phagocyte) and other specialized cells. Foremost among these specialized cells are leukocytes, whose characteristic feeding behavior was observed by German bacteriologists as early as the 1870s.[3] Pac-Man was a friendly single cell, and the Space Invaders basically colored dots on a poorly resolved screen. But they established a way of seeing; or rather, they repackaged a familiar (social)-Darwinist idea of bacteriology for the computer era: the life-and-death struggle between foreign invaders and our own "defenses."

So, too, with many of today's technologically incomparably more advanced video and computer games. Among other—

frequently "medieval"—plots, these games include fantasy variations on the city as a body invaded by extraterrestrial flying objects or microbes with a lust for high-tech war and the ultimate battle between good and evil. One example is the game *Parasite Eve,* in which a killer virus turns up in New York. It is the mutant microorganism "Mitochondria," whose origins can be traced "back to an African woman called Eve."[4] This narrative is hardly new. It limply jumbles together myriad deep-rooted images that have haunted western imagination from the nineteenth century to today (not to mention the fact that death and sin existed long before virtual Eve arrived on the scene).

Mitochondria are cellular organelles that are passed to the developing organism along with the cytoplasm from the mother's egg. Their importance in controlling the production of proteins encoded by DNA was long overlooked.[5] Naturally, like mitochondria, the "Eve" of this game is fairly active—small wonder that she can mutate and turn evil. Moreover, this "African woman" is dangerously erotic, seductive and deadly to white men, a notional holdover from the nineteenth century that enjoyed a revival amid speculation that the "African source" of AIDS somehow involved black women copulating with monkeys in the humid shadow of the jungle.[6] Hence the compelling question now raised by the computer game: "Does this micro-organism have the intention to destroy the entire human race and become the dominating species?"[7]

Darwinia, launched in 2005, was created along the same lines as *Parasite Eve,* only in a more futuristic way: "The world of Darwinia is a virtual themepark," announces the Darwinia homepage, "running inside a computer network built by a computer genius named Dr. Sepulveda. Darwinia is populated by a sentient evolving life form called the Darwinians. They are the product of a decade's worth of research into genetic algorithms." But now this evolving life is threatened by "an evil red Viral Infection. This Virus has multiplied out of control and must be stopped. Your task is to destroy the Viral Infection and save the Darwinians from extinction."[8]

The science and technology historian Timothy Lenoir has carefully shown how in the United States such games develop in close personal, but also sometimes conceptual and technical, interaction among the entertainment industry, leading universities, and military research laboratories.[9] In the process, the lines between military "reality" and pure "play" long ago became blurred. This confusion is obvious in the game *America's Army,* a tactical shooter launched in 2002 that is a recruiting tool for the U.S. Army and that in particular involves fighting terrorists armed with Russian-made weapons.[10] Indeed, the army uses game technologies in simulating the hunt for terrorists. On December 6, 2002, under the headline "Outgaming Osama," the *Washington Post* reported that a Pentagon-sponsored group was analyzing the technology behind massively multiplayer online games, played primarily in South Korea, where penetration of high-speed Internet connections outranks all other countries. The design of games that have a hundred thousand or more simultaneous online players and the technologies underpinning spontaneously arising computer networks reveal complexities relevant to the behavior and control of terrorist networks.[11]

Noting a similar connection between play and reality in the form of a computer program about simulating terrorist attacks, an article in the *San Francisco Chronicle* of August 19, 2002, reported: "It could be a scenario out of the popular video game SimCity, but updated to reflect 9/11." The program is called Weapons of Mass Destruction Decision Analysis Center and was developed by the Sandia National Laboratories in Livermore to train "war room" decision makers in the case of a terrorist attack. "The closest analogy is a war game," said John Vitko, Sandia's director, but "that's not meant to take away from the seriousness. It's very much a teaching and learning tool."

Sandia began to develop the program in early 2001, and then reality caught up: "The rash of anthrax-related deaths late last year further underscored the importance of the program, especially in the light of growing fears of bioterrorism."[12] The game industry also developed bioterrorism software for those not in a war room but at home sitting in front of a computer screen or in a video arcade. Examples include Capcom's release of the *Biohazard* shooter titles series of the popular Resident Evil games, launched in June 2001 with *Gun Survivor 2*, and the classic *Command & Conquer* series, whose 2003 version of *Generals* features the terrorist "Global Liberation Army" with a "bomb truck" that is both car bomb and bioweapon: "The fanatic at the wheel drives the truck in an enemy unit or structure to detonate the bomb on-board. Upgrades add more explosiveness or biological effects and can be used in combination."[13]

"Welcome to the Desert of the Real"

September 11, however, was no game.[14] Sometimes things happen that we—like the French psychoanalyst Jacques Lacan—can only call "the incursion of the real" in our world.

When passenger jets smash into skyscrapers, this incursion of reality assumes proportions of untold horror. For those in and around the twin towers and at the Pentagon facing death, the fear was traumatic, leaving behind a "black hole"[15] that compounded the physical wounds. There are no words to describe these wounds. For the victims, wrote Jacques Derrida, "we cannot have but a limitless compassion."[16] Nor does the trauma refer only to the invisible and ineffable wounds of the victims; the event also affected those who barely escaped as well as eyewitnesses; it even left TV viewers literally speechless. The gap arising from the inability to articulate the varieties of this trauma cannot be denied. It may be short-lived, or it may persist, making it difficult for a person to talk about what has happened. But whatever the case, the question arises whether this discontinuity brings a new form of awareness.

Slajov Žižek hoped, for example, that the attack might give "the U.S.A. . . . the opportunity to realize what kind of a world it was part of." An illusory hope, as Žižek acknowledged: "It did not; instead it opted to reassert its traditional ideological commitments."[17] That was no coincidence, inasmuch as talking about things that disrupt the symbolic order must nevertheless begin with the words and images at hand. Tradition, said Marx, is the "nightmare" that "weighs on the brain of the living." The "nightmare" of inherited words and images quickly takes speech back into familiar territory. Not that we cannot occasionally say something new with hoary language and stock images. But old "names, battle slogans, and costumes"—Marx's metaphors for customary ways of thinking—work to prevent events like 9/11 from opening a window onto the new.[18] According to linguist Sandra Silberstein, we should have no illusions about how quickly old patterns of speech were employed to "understand" the terrorist strike on the two towers. Indeed, from the outset, the performance of the media itself with its "carefully crafted rhetoric and imagery" was a major factor in "building a consensus around war."[19] CNN's live reporting in the first few hours carefully stuck to the rules of infotainment, for instance, when reporters systematically asked some survivors about blood and screaming on the stairways.[20] This tactic became particularly clear when astounded or disgusted interviewees refused to answer such "leading questions."

Silberstein argues further that the most significant performative act particularly of the nonprint media consisted in creating a new demarcation between "them" and "us," and at the same time establishing the president as commander-in-chief in a new war. She cites the example of one of the first

CNN interviews in which, following a detailed account of her escape from the north tower, a woman says of the attackers: "I just hope Bush will do whatever is necessary to get rid of them."[21] It is also well known how quickly and effectively public tears, the ostentatious display of the American flag, and widespread profession of faith and prayer over American airwaves helped to fill in the "gap" (Žižek) between experience and language. As a counterexample, Žižek cites an interview broadcast only once—in typical CNN style—of a fireman's widow who confessed that she did not believe in the "strength" of prayer and didn't think much of military retaliation.[22]

But there is an entirely different set of words and images suited to symbolizing the experience. For a long time popular culture has been producing images that prepared our media-sensitized ways of seeing specifically for September 11. In some cases it went as far as the event not even being perceived as such. During the live transmission of the second plane attack on the WTC, television stations in Europe received calls not from distressed viewers but from viewers enraged that Hollywood disaster films were now even being broadcast in the afternoon. On September 11, Swiss TV was forced to drop from the evening schedule a film that would, however, have been entirely fitting: Tim Burton's comedy *Mars Attacks* of 1996. Associating Hollywood movie images with the September 11 attacks has become commonplace. Michael Bays' *Armageddon* (1998), Roland Emmerich's *Independence Day* (1996) and *Godzilla* (1998), and the male fantasy of the destruction of skyscrapers in San Francisco's financial district in David Fincher's *Fight Club* (1999) prove that these representations of destruction are part of our mass-media-shaped imagination.

The TV images from September 11 cannot really have surprised us; image and experience converged even as far as semantic details. In *Fight Club,* as the end—the planned (imaginary? real?) blowup of the skyscrapers in the financial district—looms, the protagonist Tylor Durden says quietly, "Three minutes. This is it. The beginning. Ground zero."[23] *Ground Zero:* a term from the nuclear test era denoting the site of a bomb blast left denuded, burnt, eviscerated, and radioactive. The fantasy of collapsing skyscrapers, as depicted in *Fight Club,* becomes a weapon of mass destruction as devastating as any atomic bomb.

But to get back to the airplanes: Consider the students responsible for the Columbine High School massacre. They, too, planned to hijack an airplane and destroy the Empire

State Building, carefully teaching themselves sharpshooting using videogames.[24] One might ask whether the impact of the American Airlines jets on the two towers was so deeply unsettling not because it represented something far from outrageous or unthinkable but rather because it was something "frightening which leads back to what is known of old and long familiar." Freud observes that "an uncanny effect is often and easily produced when the distinction between imagination and reality is effaced, as when something that we have hitherto regarded as imaginary appears before us in reality."[25] Jean Baudrillard caused a scandal not only by interpreting September 11 in precisely this way but also by adding that in this sense the towers had committed "suicide."

For him the attack on the towers was the realization of a taboo desire in the classical Freudian sense: "They *did it*, but we *wished for* it." Then, according to Baudrillard, "no one can avoid dreaming of the destruction of any power that has become hegemonic to this degree" but that is "unacceptable to the Western moral conscience."[26] And this "they" should not be misconstrued. For Baudrillard "they" are no longer Islamic others but simply the counterpart of a western—American—system of dominance so total that those who have paid for the system are now "chang[ing] the rules of the game" and staking their own lives by injecting a singularity—a definitive act—into the heart of the global exchange circuit.[27]

The Persian-German Islamicist and journalist Navid Kermani has both clarified and sharpened this rather general hypothesis—and I do not say that it is wrong—in a very illuminating way.[28] For Baudrillard, suicide tacitly remains an option for an "other" that he leaves unexplored but that in the end somehow seems "Islamic." Baudrillard also proceeds on the assumption that terrorists "hide" behind the screen of modern, everyday American life. But Kermani bluntly questions this very widespread and readily accepted idea. He understands the wish to destroy the towers as "our" wish, but he deciphers the act itself, airplane pilot suicide—akin to a living videobomb—as a fundamentally western pattern of behavior.

As Kermani shows, Shiite Islam does have a certain tradition of religiously motivated martyrdom ritually acted out through folk religion in passion plays and parades of flagellants. This piousness celebrates the self-sacrifice of the prophet's grandchild, Imam Hussein ibn Ali, in the year 680 at Karbala. The event marks the beginning of the Shiite faith, which today finds expression as "a striking altruism that especially impresses European visitors." At the fringes of this faith the figure of the suicide aggressor—for instance, the child soldiers in the Iraq-Iran war—may have a place. But the crucial point is this: "The cult of martyrdom is clearly a Shiite phenomenon which, in the first instance, only developed in opposition to the Islamic majority . . . By contrast, the ideology of the terrorists, as far as we know anything about it, is definitely Sunni." For Sunnis like Mohammed Atta, the act of September 11—the suicidal martyrdom—belongs to beliefs that "they regard as heretic and which in turn reflect Zoroastrian and possibly also Catholic origins."[29]

Here the label "Islamic" or "Islamist" explains little or nothing. So Kermani inverts the western view that the difference between Sunni and Shiite is minor and focuses it on the west itself. According to Kermani, the attackers are members of the westernized middle and upper classes of Egypt and Saudi Arabia who in Germany and the United States were completely integrated and led a westernized life. They went to the disco, enjoyed Hollywood disaster films in the evenings, and got high on the weekends. They had nothing to hide because that was their life. But like many other young men and women both west and east, north and south, they came under the spell of religious fundamentalism that—and this is equally true of Islamic fundamentalism—is a phenomenon of modernity, syncretic, a mix of old and new, of western and Islamic images and beliefs. The French Islamicist Olivier Roy argues similarly: "The real genesis of Al Qaeda violence has more to do with a Western tradition of individual and pessimistic revolt for an elusive ideal world than with the Koranic conception of martyrdom."[30]

Kermani shows convincingly that above all, the idea of going to one's death as a living missile—a "pure deed," without declaration or demands or conditions—is perhaps the most western, indeed the most modern, element of Islamic fundamentalism. This nihilistic attitude has its roots in Nietzsche and in German romanticism, to which, incidentally, Japanese Kamikaze pilots also made fond allusion. It reached its highest expression in the ideas of Ernst Jünger and the German National Socialists, to the point where, in the Second World War, they planned to build suicide planes (the Japanese beat them to it). But in no way does it relate to the religious thinking of Sunni Islam.[31]

So, a double, even multiple familiarity. Not only were the perpetrators of September 11 conscious of having drawn on popular culture in conceiving their attack and could prepare in the open because they were living a life that was entirely "theirs."[32] Even their deed is much more in line with western tradition than with "Islam," leading from the Christian cults of sacrificial death and martyrdom through the German romantics to Nietzsche and the western nihilism of the twentieth century. To recapitulate: For both victims and spectators (who are more familiar to us), even—or in particular—the crash of commercial jets into skyscrapers followed de facto the rules of a crash anticipated long ago and did not simply register as a shocking event.

Considering the already familiar images and representations of such deeds, Slavoj Žižek inverts the relationship between "image" and "reality" in drawing the following conclusion: "What happened on September 11 was that this fantastic screen apparition entered our reality. It is not that reality entered our image: the image entered and shattered our reality (i.e. the symbolic coordinates which determine what we experience as reality)."[33]

Anthrax!

How confusing. What exactly was it that shattered our (television) reality? Images? The real? Or simply terror (the public opinion favorite)? When we talk about anthrax now, the first answer is easy: it was the mail. On September 18, 2001, the first three letters containing anthrax spores were sent to NBC headquarters, the *New York Post,* and the *National Enquirer* at American Media Inc. (AMI) in Boca Raton, Florida. On October 4, the media reported the first case of pulmonary anthrax: British photo editor Bob Stevens, who worked at the AMI building in Boca Raton, had inhaled powder from a letter addressed to "Jennifer Lopez." Stevens was admitted to the hospital on October 2 and died on October 5 of pulmonary anthrax. When anthrax spores were found around the photographer's desk the next day and another colleague became ill, the Federal Bureau of Investigation was called in.

On October 7 the United States launched its initial air strike on Afghanistan; two days later, the U.S. media raised the specter of "bioterrorism." That same day, October 9, two anthrax letters were sent to Thomas Daschle and Patrick Leahy, senior Democratic party leaders in Congress. On October 10, another colleague of Stevens tested positive for anthrax; suspicion now focused firmly on letters (but the one sent to Stevens could not be found). On October 12, the media reported the illness of one of Tom Brokaw's co-workers at NBC, and in an interview Vice President Cheney voiced the suspicion that Al Qaeda was behind the letters. On October 15, the letter addressed to Daschle was opened in his office. The anthrax spores in the two letters to Daschle and Leahy turned out incidentally to be more aggressive, more finely milled, and therefore much more dangerous than those found in the letters sent to the media.

On October 16, Washington, DC, postal workers Thomas Morris, 55, and Joseph Curseen, 47, showed symptoms of pulmonary anthrax; on October 25, they died. Both worked at the Brentwood Mail Center and were likely infected by the letter sent to Senator Leahy. On October 31, mostly likely spores from the letters to the New York media killed postal worker Kathy Nguyen, 61. Finally, 94-year-old Ottilie Lundgren succumbed in Oxford, Connecticut, on November 23, following accidental cross-contamination of a mail sorting machine.[34]

The case seems clear-cut. The images of people wearing biohazard suits that appeared in the worldwide press and on television during and after the event show perhaps more pointedly than all the pictures of toppling skyscrapers what we *really* imagined to be the greatest danger and deadly menace. Ever since the German Oberste Heeresleitung (Supreme Army Command) ordered the use of novel "gas weapons" as part of their so-called disinfection operation in the battle of Ypres on April 22, 1915, invisible agents that contaminate the air and breathing itself represent the quintessential modern threat.[35] All the same, it is hard to countenance Peter Sloterdijk's view that terror originated with the use of gas weapons: ever since the time of the French Revolution, "terror" has described violent, relatively low risk threats and practices whose unforeseeable and "unfair" use produces the maximum possible dread and intimidation.[36]

What the pictures of men in gas masks from the First World War do suggest is the source of modern perceptions of threat. The shock that the poison gas provoked in soldiers already accustomed to the enormous and unprecedented horror of trench warfare signaled a technological break with the past. And the various chemical weapons were only the most obvious form of modern invisible threats; for in parallel, German World War I agents had already starting using anthrax spores on Allied draft animals, cavalry, and food animals to disrupt troop movements and supplies.

Although anthrax is the bioweapon par excellence, for a long time chemical warfare agents figured more largely. In World War I various forms of poison gas killed some 90,000 soldiers; roughly a million more were injured by acid gas, some seriously. In 1920–21, British troops invading Iraq used mustard gas; the Italians used chemical warfare agents in Libya, and the Spaniards in fighting Berber tribes in Morocco;

in 1936 Italian troops again turned to mustard gas in the campaign against Ethiopia. Although poison gas disappeared from the European theater of war after 1918—not counting the use of Zyklon B in the German extermination camps from 1942 to 1945—in the colonial wars and in the Near East they became a tool of European power.

In World War II, chemical and biological weapons were not deployed as instruments of war because their tactical and strategic impact in a battle with equals—unlike in the colonial wars—would be minimal. Only after the war did it come to light that the Japanese army, like the Waffen SS during the war, systematically conducted trials with bacterial weapons on prisoners of war. These weapons had allegedly been deployed by the Japanese here and there. Intelligence agency reports that Tokyo and Berlin possessed bacteriological weapons led the United States, as early as 1943, to take over the former military base in Fort Detrick, Maryland, for the development of bioweapons: anthrax agents aimed at soldiers, and agricultural pests targeted to Japanese rice and German potatoes.[37]

Following the war, in the shadow of the atom bomb, a secret arms race began to develop weapons systems with biological agents on both sides of the ideological front lines. No one knows whether these weapons, which were outlawed internationally in 1972, would ever have been used by the superpowers and their allies or client states. What we do know is that 1945 marked the beginning of the era of suspicion. In 1948, Israeli soldiers were claimed to have poisoned Arab wells; during the Korean War, North Korea and China accused the Americans of bacteriological warfare; since the 1970s, the Cubans have repeatedly asserted that the United States intends to poison the island's agriculture; finally, in 1980, the United States accused the Soviets of using chemical weapons against Afghanistan. At the very least, the record attests to the U.S. Air Force's use of the herbicide Agent Orange in defoliating forests and destroying Vietnamese plantations, the deployment of chemical warfare during the Iraq-Iran War, and gas attacks on Iraqi Kurds by Saddam Hussein's army in 1988.

This history, whose most important stages I have only touched on here, gives reason to fear chemical and biological agents. But not only. The terror dating from the end of World War II, which can be summarized under the tidy acronym NBC (nuclear, biological, chemical weapons), also feeds on assumptions and pure suspicion, on speculation, and on fear of contamination (haven't the Jews always poisoned wells

. . . ?).[38] And it feeds, too, both particularly and generally on the idea of an invisible and thus scarcely comprehensible threat. The history of bioweapons and "bioterror," as people began to say in the 1990s, is as much a history of facts as it is a narrative, an explanatory model, built on the phantasms that I will examine in this essay.

To get back to September 11: Does this model also apply to anthrax? That is, does it apply to bacteria that, outside of a few high-security laboratories, are not supposed to exist but that in autumn 2001 suddenly made their appearance in our everyday life as a lethal danger in simple envelopes? It seemed as unbelievable as passenger jets flying into skyscrapers—yet real. A fact, not a story. In the case of anthrax, one might point out, it wasn't actually an "image" that shattered our reality, as Žižek writes, but completely the opposite: an invisible yet deadly substance.

There is no doubt that anthrax can be dangerous, even fatal. Antibiotics notwithstanding, the illness triggered by the bacterium is life-threatening. The bacterium itself was isolated in 1876 by Robert Koch, who also photographed it.[39] Louis Pasteur subsequently diluted it in the laboratory. His efforts were so successful that in 1881, in a famous inoculation experiment at Pouilly-le-Fort, he was able for the first time to save twenty-four sheep from the dreaded disease.[40]

Pasteur's and Koch's anthrax appears nevertheless to have been a different sort than the one in the five letters—even beyond the obvious differences between the bacteria bred in cell cultures around 1880 and the so-called weapons-grade bacteria of today. Pasteur and Koch wore no special clothing. They did not work in a high-security facility but in a relatively primitively outfitted laboratory in the Ecole Normale Supérieur on the rue d'Ulm in the center of Paris (Pasteur) and the back room of Koch's country doctor practice in Wollstein. At most they would have washed their hands after handling the anthrax. Protective masks and biosafety air cleaners were unthinkable. The windows in Paris and Wollstein had ordinary, wood-framed panes, and the gentlemen would have clearly rebuffed the impertinence to wear rubber masks ruffling their carefully clipped beards.

This historical reminiscence is just to suggest that it is not always very clear what exactly is being said in inflated media discourse when the subject is "anthrax," as it was in autumn 2001. The denotative sign "Warning: Anthrax!" seems to say clearly what it means. And yet it is deceptive. Anthrax is not always synonymous with anthrax, and especially not with "anthrax." And that of course raises the question how dangerous anthrax "really" is.

So first we have to beware of a distinction that today separates "our" anthrax from the anthrax of "the others." One might formulate the following rule: "Anthrax, caused by the spore-forming bacterium *Bacillus anthracis,* is rarely seen in industrial nations but is common in developing countries." The authors of this statement, doctors at the Yuzuncu Yil University School of Medicine in Van in eastern Turkey, reported in June 2001 an annual incidence of two dozen cases of cutaneous anthrax, which is endemic in Turkey.[41] The same could be said for Russia, Kazakhstan, India, the Near East, and several African countries. But the World Health Organization's (WHO) *World Anthrax Data Site* also counts Spain among the countries with dozens of annual cases of human anthrax.[42] In North America, anthrax regularly appears as livestock outbreaks; in several counties in Texas it is endemic among cattle. But human infection with cutaneous anthrax is also a frequent occurrence in the United States—for example, in August 2000 in Minnesota or in the last case before September 11, near Rock Springs, Texas, in July 2001.[43] Cutaneous anthrax arises when the spores enter a wound; it is easier to cure than pulmonary anthrax, in which a fair number of spores have to enter the lung before a person becomes ill.

Even accounting for such differences, anthrax never interested "us" before the letters. We knew nothing about these cases of anthrax, and didn't have to. A person could vacation in Spain—or be governor of Texas—without having to think about anthrax in the slightest. And so for the people of Turkey (which broadcasts no global press conferences when people with anthrax go into the hospital) and Spain (where according to WHO the statistics are a little slipshod) anthrax was obviously less dangerous than it became for "us." Dangerous anthrax became a "reality" in autumn 2001—without losing its paradoxical nature as a bacterium that is both widespread and restricted to high-security laboratories.[44] The spores found on Capitol Hill in October 2001 were real and could kill people—but exactly how "real"? So real that the building they appeared in (the House of Representatives, Republican majority) had to be evacuated? Or, as the media reported at the time, just real enough that politicians (the Senate, Democratic majority) could carry on working in peace at least a while? And in ways both enigmatic and banal, this ambivalent assessment of how dangerous anthrax "really" is is reflected in the texts of the four anthrax letters that were sent on September 18 and October 9. The brief September 18 letters to NBC's Tom Brokaw and the editor of the *New York Post* begin with the words:

> This is next.
> Take Penacilin [sic] now.[45]

The October 9 letter to Senator Daschle (a similar letter went to Senator Leahy) reads:

> You can not stop us.
> We have this Anthrax.
> You die know [sic].
> Are you afraid?

Cynical and contemptuous they may be, but these letters are interesting. On the one hand, it is clear that they are warning the addressees—the intent is not covert assassination but the message: We have "it"; we can do "it." On the other hand, the letters announce the danger of anthrax each in a different way, and thereby teach us, incidentally and unintentionally, an epistemological lesson: even so obvious a pathogenic agent as *Bacillus anthracis* can be differentially designed as a biological entity—totally deadly, or a threat to keep in check with antibiotics.

And that takes us to the second aspect of the obvious inconsistency of the denotative sign "Warning: Anthrax!" For despite the five deaths, in autumn 2001 "anthrax" was primarily what Douglas Rushkoff calls a "media virus."[46] That becomes clear when you consider how anthrax cases were communicated and represented worldwide in fall 2001, and when you compare that representation with WHO's distribution map of worldwide anthrax cases, quoted above. The CNN map titled "Anthrax cases around the world" shows not human anthrax infections worldwide but the emergence of anthrax *letters* in individual countries. Especially remarkable is CNN's information regarding Kenya. The country sounded the anthrax alarm as early as September 2001. CNN notes: "A suspicious letter mailed on September 8 from Atlanta, Georgia, to Nairobi, which originally tested positive for anthrax, now tests negative, according to the Kenyan Health Ministry. This was the first confirmed case of anthrax outside the United States."[47]

That is the empirical proof, if you will: Anthrax is caused not solely by bacteria or the right spores but also by letters. This point is made beautifully on a PBS companion Web page to a NOVA program on bioterror. Clicking "Bioterror" on the Web page gets you detailed information about various disease agents, and among the corresponding illustrations of the pathogenic organisms, the one for anthrax shows nothing other than—letters. For the agent of anthrax *is* a letter, in the same way that *Vibrio colerae* (isolated by Robert Koch in 1884) is the agent of cholera. To put it more precisely: Producing imaginary "anthrax" requires a bit of contamination, a tiny invisible speck of the real in a world of signs and images. But, as I said, it also requires much more and especially: letters.[48]

Whoever circulated the anthrax letters did his or her homework in postmodern media theory. The five poisoned letters were specifically aimed at media people and politicians because the letters needed to end up where anthrax spores could transform into imaginary "anthrax" directly, quickly, compellingly, and globally. The microbiologist Barbara Hatch Rosenberg of the Federation of American Scientists surmises that even in advance of media reports on the anthrax cases themselves, hoax letters—that is, letters not containing spores—were also dispatched to the media, to enhance the effect of the attack.[49] This theory is of course controversial. But it makes an important, if apparently trivial, point: that there can be no social reality outside of the media, as the perpetrator or perpetrators knew very well. Social reality is what we perceive, and our perceptual apparatus—a network of neurons, eyes, ears, symbolic systems, news agencies, electronic media, paper, and letters of the alphabet—was only ready to "see" anthrax in autumn 2001, under very specific circumstances.

That explains why it proved so difficult to distinguish the illusion from reality. As everybody knows, after the few letters from the perpetrator, all further "cases" of anthrax were pure fakes, just hoax letters, nothing but imitations of the letters shown on TV—part of the media feedback loop. In the United States alone, after October 1 the FBI pursued around 2,300 cases of alleged "anthrax" letters.[50] For all the other instances, for example, many in Europe, I could obtain no figures; but they also must certainly amount to thousands. "Anthrax" hoax letters were hardly new: in the two years before September 11, U.S. authorities counted 178 "anthrax" hoax letters targeted at various government agencies. Even the media were so accustomed to them that in autumn 2001 they threw the letters containing the actual anthrax spores in the wastebasket (where they were retrieved).[51]

The history of the "anthrax" hoax letters is well documented, in particular the spate of such events from October 1998 to February 1999, and most notably in California, aimed at abortion clinics, but also, as I said, at authorities and next at the media and schools.[52] But this history is not easy to interpret. Did these hoax letters have something to do with the fear of bioterrorism that was growing in America at the time? Or ought they to be understood simply as a temporary means of intimidation on the part of right-wing extremists? Is there a connection between these events and the "anthrax" letters that followed September 11? There is some evidence for such a connection: 550-odd hoax letters were also sent to abortion clinics in fall 2001; the right-wing group Army of God claimed responsibility for them.[53] Generally speaking, however, the hoax letters have been too little investigated to discern a pattern. What is striking is that most of the senders of the hoax letters were identified and arrested, and that these letters never contained real spores. In any event, the hoax letters had an effect thanks again to the media's detailed reporting in each case. Indeed, whereas in 1998 and 1999 such letters were largely a Californian phenomenon, after September 11, their impact became global.

In the context of the terrorist attacks, the media had no compunction about applying all of their new technological wherewithal. The hardware of the worldwide media network—especially the massively increased capacity of the global data highway in the last century—facilitated transmission of all those pictures of postal workers, contaminated letters, and men in gas masks, thereby contributing to an ever-growing global media construction of "anthrax." In the end, even in Switzerland mail centers had to shut down, and ballots from a cantonal referendum in Lausanne had to be counted using rubber gloves and face masks. As never before, the contaminated letters and all the circulating uncontaminated copies created an imaginary space in which fear was disconnected from its concrete object, cloned itself, and hypertrophied. "Anthrax" was highly contagious and dangerous. Whereas pathogenic anthrax killed five people, metaphorical "anthrax" poisoned the imagination of millions.

"Anthrax" became a part of September 11 because it was the perfect terror: not only do "they" attack skyscrapers, they *poison* us—everywhere, constantly, at random. President Bush gave fitting verbal form to this diffuse feeling when he stated at a press conference on November 7, 2001: "We fight a new kind of war. Never would we dream that someone would use our own airplanes to attack us and/or the mail to attack us."[54] We will see that neither is true, in other words, that both attacks with airplanes and with letters had long been expected.

2

Bioterror and Weapons of Mass Destruction

That the "anthrax" letters unleashed a hysterical wave of fear, totally out of proportion to the actual threat, is not news. All the same, it is worth recalling that the imaginary, too, which is the subject here, proliferates in a space that is controlled by the *letter.* "Letter" can mean both correspondence and constituents of the alphabet. And the anthrax letters are a textbook example of how this *letter*—known in the parlance as the circulating signifier—is the real agent of social reality. This point is key. In other words, we need to understand how the signifier "anthrax" spread about and linked up with other signifiers, whose scope and impact then became global and "strategic." I want to show how, starting with the anthrax letters, the September 11 terrorist crash of commercial jets transmogrified into proof of a threat posed by weapons of mass destruction, masterminded by none other than Iraq. Biological anthrax became metaphorical "anthrax," with its connotations of Al Qaeda, bioterror, WMD, Iraq, and so on. We will see that the top levels of the Bush Administration desired the Iraq War even before 9/11, which raises the question whether the letters also were "desired." I will come back to this point.

Signifier/signified

Analyzing and understanding the impact of the anthrax letters is helped by knowing a little theory from the field of cultural studies, specifically, poststructuralist textual theory. In the Prologue, I quoted a statement by Jacques Lacan, namely, that metaphor—"say[ing] something by saying something else"—is the "core" of language because it is language at its most effective.[1] Lacan had in mind the language theory of Ferdinand de Saussure (1857–1913) as taken up by structuralism and reformulated by (in particular) Lacan himself and Jacques Derrida.

De Saussure introduced two important terms to describe very precisely what he called "a sign": the "signifier" (*signifiant,* the tangible, phonetic, or graphical unit of meaning) and the "signified" (*signifié,* what is meant). Lacan adapted Saussure's terms in a way that proved highly influential, and he represented them graphically as follows: $\frac{S}{s}$. This particular arrangement illustrates the domination of the signified by the signifier, and the separation of both by a "bar."[2] Technically speaking, the signified—or meanings—can slip and slide under the bar; that is, word meanings change more or less significantly based on context. According to Lacan and Derrida, language basically functions according to shifting patterns of the age-old rhetorical figures of metaphor and metonomy. A metaphor, says Lacan, is "one image for another" (the small "s" under the bar is exchanged), and metonymy "one word for another" (the large "S" over the bar is exchanged, and with that another small "s" comes into play). Put another way, language never denotes things in exact one-to-one correspondence; rather, it paraphrases them, compares them, and invents pictures for them. (Only in mathematics can one hope to find a definite, precise expression for a thing.)[3] There is no "fixed" or "natural" connection between signifier and signified: words are fundamentally and always equivocal. Meanings can vary and shift, an idea Derrida drove home. The conventional, literal meaning of a word—which is obviously essential for communication—is only *one* possibility, one alternative among all the possible meaning effects of a word in specific contexts.

In this chapter we will see that such an analysis can help us to understand the many meanings of anthrax/"anthrax" as well as the words associated with this signifier. For the signifier can mean not only "anthrax" the bacteria or the illness released by the bacteria but also entirely different things. The underlying premise is that the signifier *as such* is not yet inherently "meaningful" but "stupid" (Lacan), an "empty" form to be filled by alternating signifieds—"content," "meanings" (among which is the meaning that in discourse becomes fixed as the conventional, "actual" meaning). Members of the heavy-metal band Anthrax were the first to make this particular connection when in October 2001 they addressed the public and unconsciously illustrated signifier theory: "In the twenty years we've been known as 'Anthrax,' we never thought the day would come that our name would actually mean what it really means"—that is, the conventional meaning. "When I learned about anthrax in my senior year biology class, I thought the name sounded 'metal' . . . 'Anthrax' sounded cool, aggressive"—that is, the metaphorical meaning—"and nobody knew what it was . . . To us, and to millions of people, it is just a name"—that is, just a signifier.[4]

But not everyone saw it that way, and the climate of thinking changed suddenly. For the physicians of Yuzuncu Yil University in Turkey, "anthrax" had long designated the skin rash common among their patients as well as the difference between industrialized and developing countries. But in autumn 2001, in the five poisoned letters and concomitantly in the frightened perception of the media-sensitized public, "anthrax" signified "Death to America," or "Allah is great." The connection seemed clear: It was as if, beginning with President Bush's speech before Congress of September 20, 2001, officially declaring Al Qaeda to be the arch enemy of the United States in the "war on terror," the anthrax letters could have had but one possible source—even though at the time Bush made no mention of anthrax.[5] Very quickly, "anthrax" came especially to mean Bin Laden and the Taliban. A bogus Anthrax album cover circulated over the Internet showing the musicians dressed up like the Taliban with flowing beards and the lead singer as Bin Laden—"armed and dangerous, coming to a city near you"—and likewise a fake Uncle Ben's rice box cover with "Uncle Bin" plumping for "Instant Anthrax" ("Fortified—Enriched—Powdered").

This wasn't just tomfoolery on the Web but reflective of the perception that the signifier "anthrax" had been building since October 2001. That much is evident in a billboard created by the U.S. Department of Defense in autumn/winter 2001 for members of the armed forces displaying strains of *Bacillus anthracis,* and—in odd contrast to the administration's official denials—medical syringes morphing into falling bombs. Here it becomes clear that vaccination against "anthrax" is bomb warfare: under the category "war on terror," the anthrax letter terror constitutes "bioterror," and thus is naturally connected with the Taliban in Afghanistan (who must be bombed) and with Al Qaeda.

Such are the metaphorical effects of the signifier anthrax. Since September 11, "anthrax" was generally assumed to refer to the author of the biological letter bombs—Al Qaeda. Yet this remained a very vague assumption. Moreover, the scare word "anthrax" quickly ceased to mean primarily terrorist "others" when it became clear by the end of 2001 that the perpetrator or perpetrators had to be a close insider of the American bioweapons research community. I will come back to this point.

So: from November 2001, and even more following January 2002, "anthrax" initially was metonymically supplemented by two signifiers—"bioterror" and "weapons of mass destruction"—and then was pretty much replaced by them. That had the double advantage of suppressing the concrete, unexplained anthrax story and at the same time more effectively reinforcing the awareness of it. "Bioterror" no longer simply meant "anthrax" but basically everything evoked by the dark, primordial fears that people have always associated with highly contagious diseases such as cholera, smallpox, and black death, or exotic viruses such as Ebola and Marburg, which cause their victims to bleed to death.[6]

In 2002 these were not the only fears, though. The specific coupling of the two signifiers "bioterror" and "weapons of mass destruction" was becoming increasingly virulent. This link-up had been more or less present in the American consciousness since 1996 (as we will see), but up to then it had not played as significant a role as it was now being pressed to do. Post-anthrax, the signifier "WMD" began to suggest, however imprecisely, that biological weapons can spread not just fear and horror—terror—but that the possibility actually exists to kill thousands of people at a single blow, as happened on September 11.

Weapons of Mass Destruction

The counterarguments are obvious: The alarming connection between bioterror and weapons of mass destruction is justified by the facts. An incalculable number of people would be at risk of dying if genetically modified, possibly novel microorganisms were deliberately released. These agents are the diabolical "gift that keeps on giving," as Bill Clinton so cogently remarked.[7] Just after September 11, Slavoj Žižek, too, noted that "the true long-term threat is further acts of mass terror in comparison with which the memory of the WTC collapse will pale—acts that are less spectacular, but much more horrifying. What about bacteriological warfare, what about the use of lethal gas, what about the prospect of DNA terrorism (developing poisons which will affect only people who share a specific genome)?"[8] So aren't there actually more than enough reasons for genuine worry? After all, it wasn't unthinkable that politically unpredictable regimes like Saddam Hussein's could supply themselves with bioweapons. In 1995, in exchange for the U.N. "Oil-for-Food" program, the Iraqis admitted that they had such programs and grudgingly accepted destruction of their NBC (Nuclear, Biological, and Chemical) production facilities by U.N. inspectors—to what extent remained controversial right up to the invasion of Baghdad by U.S. troops.[9] And isn't it possible that other terrorist groups like Al Qaeda were involved with these materials, as the intelligence agencies have repeatedly implied?[10]

Ever since the waning of the threat of nuclear war by the former Soviet Union, a new horseman of the apocalypse appeared on the far horizon of military planning: a sepulchral vision of mass annihilation through biological weapons. *New York Times* journalists Judith Miller, Stephen Engelberg, and William Broad delved into the subject in depth. They reveal that in the final years under Gorbachev, the Soviet Union still had a huge production capacity for advanced biological agents. During and after the fall of the Soviet Union, a variety of American government agencies strove in cooperation with Soviet scientists and bioweapons experts either to destroy plants used to produce biological agents or to convert them to civilian production. The funds required particularly for the latter were granted by the U.S. Congress, but only in very limited amounts.

The American experts cited by Miller, Engelberg, and Broad were seriously worried, especially by the many unemployed Soviet former bioweapons experts: according to an oft-voiced apprehension, they could be recruited by terrorist groups or "rogue states"—a term introduced by the Clinton administration in 1993. As a very general response to this threat, in 1996 the Clinton Congress passed the Defense against Weapons of Mass Destruction Act, dealing with the development of nuclear, chemical, and biological weapons in specific states and transfer of these weapons to "terrorists."[11] This so-called Star Wars law revived an old idea of Ronald Reagan, which was to protect the United States from approaching nuclear rockets by constructing a space-based antimissile defense screen. With the end of the Soviet Union, the same danger was projected, so to speak, onto new potential instigators: rogue states that could have acquired the technology and personnel skilled in the ballistic weapons of mass destruction of the fallen superpower.

According to this scenario, the weapons would not simply be conventional nuclear warheads but also warheads equipped with bioweapons. In the first half of the 1990s, the suspicion was finally raised that terrorist groups could also obtain such weapons and threaten the last remaining superpower with extinction. At the time, the suspicion remained a vague bit of rhetoric with only the barest whiff of political significance. But the future phantasm had already begun to take shape.

This threat scenario cannot be rejected out of hand. But even if it is theoretically possible, to date there is no evidence for such remote consequences of the former Soviet bioweapons program. Despite claims to the contrary, there is also no proof that any terrorist group had access to anthrax, plague, or smallpox from Russian laboratories or production sites, nor, as it is feared, that there was any bioweapons brain drain.[12] Likewise, it is not true—though oft claimed—that anyone who can fake a research university letterhead can order anthrax by post from the appropriate supplier, and anyone who knows his way around the Internet can find easy-to-follow instructions for making weapons-grade anthrax, using equipment purchased at the corner hardware store.[13] Malcolm Dando recently pointed out the substantial difficulty a wannabe terrorist would have carrying out—even barely carrying out—an anthrax attack:

> Preparing an agent like *Bacillus anthracis* for use in an airborne biological weapons attack is not an easy technical task. It certainly appears to have taken years of experimentation to perfect in the state [i.e. Soviet] offensive biological weapons programmes of the last century. Such "weapons grade" material would have had a very high

> concentration of spores, uniform particle size, low electrostatic charge and would have been subject to other treatment to reduce clumping of the material. There is, however, a prior problem for the would-be attacker. Because of the natural processes of mutation and geographical isolation in different environments, there are many different strains of anthrax in various parts of the world and these have different levels of lethality for humans. The first problem for a weaponeer, therefore, would be to find a virulent strain of anthrax.[14]

If that weren't the case, after September 11 there would have been not five but vastly many more letters containing anthrax spores. Moreover, right-wing extremists in the United States who use hoax letters to repeatedly threaten hospitals would long ago have followed up their threats with action. Finally, why would Iraq go to the considerable effort and expense of importing western biochemical technology and building factories just to work on an anthrax weapons project?

I will come back to the sometimes powerful speculation over the "possibility" of such a danger; here it suffices to say that "bioterror"—and especially the discourse on WMD as it developed between September 11, 2001, and the Iraq War in

March and April 2003—worked so well precisely because it did not rely on empirical evidence. In conjuring up threatening possibilities that were all the more frightening for being left unexamined, the discourse could evolve unfettered. It was not concerned with facts, rather, as I will now show, with very effectively associating certain signifiers.

In analyzing the discursive events leading to the Iraq War, I will limit myself to observing how openly words were manipulated, how some words came to replace other words, and how evidence was produced by using the same words over and over, repeating them, and associating them with one another. I will first examine how the Iraq War was justified and "sold" by the frequent use of certain words alone and in combination. That does not constitute the whole of the story I would like to tell in this essay, but it is a start.

The $64,000 question is, How did the term "weapons of mass destruction" come into play following September 11? In his speech of September 20 President Bush was clear about the need for America to prepare itself for a long "war on terror." But at the time—that is, before the actual anthrax crisis—the signifier "weapons of mass destruction," which would become such an important instrument in the campaign, had not yet made its rhetorical appearance.[15] Then on

September 22, the online edition of *Time* magazine announced that the September 11 assassins might have tried to use a crop duster to spray bioweapons, and in the following days the anthrax letters began to reach their destinations.[16] Bush addressed the topic at a press conference on October 11: "You may remember recently there was a lot of discussion about crop dusters. We received knowledge that perhaps an al Qaeda operative was prepared to use a crop duster to spray a biological weapon or a chemical weapon on American people . . . We knew full well that in order for a crop duster to become a weapon of mass destruction would require a retrofitting." But America's reaction to this imaginary threat was hardly benign: "We talked to machine shops around where crop dusters are located. We took strong and appropriate action. And we will do so any time we receive a credible threat."[17]

This was the first time since 9/11 that Bush had broached a possible bioterror threat and in so doing used the signifier "weapons of mass destruction"—a metonym for "anthrax." In signing the Patriot Act on October 26, after its hurried passage through Congress, Bush addressed the issue again: "Current statutes deal more severely with drug-traffickers than with terrorists. That changes today. We are enacting new and harsh penalties for possession of biological weapons."[18] But the connection with weapons of mass destruction did not emerge here. Nor did Bush make such a connection during a November 3 radio address that he devoted to the anthrax attacks, appealing to the people under circumstances of grave concern.[19]

The change came soon after. In a speech to a conference on terrorism in Warsaw on November 6, Bush made explicit the metonymic shift to weapons of mass destruction and terrorism, and thus their connection: "These terrorist groups seek to destabilize entire nations and regions. They are seeking chemical, biological and nuclear weapons. Given the means, our enemies would be a threat to every nation and, eventually, to civilization itself. So we're determined to fight this evil, and fight until we're rid of it. We will not wait for the authors of mass murder to gain the weapons of mass destruction."[20] Two days later, in a joint press conference with Tony Blair, Bush stated: "But our nation and this terrorist war says to me more than ever that we need to develop defenses to protect ourselves against weapons of mass destruction that might fall in the hands of terrorist nations. If Afghanistan or if the Taliban had a weapon that was able to deliver a weapon of mass destruction, we might be talking a little different tune about our progress against al Qaeda than we are today."[21]

In a November 8 press briefing, then National Security Advisor Condoleezza Rice made the connection between terrorism, weapons of mass destruction, and Iraq.[22] And in his speech before the U.N. General Assembly on November 10, Bush spoke in an unmistakably sharper tone of the supposedly clear link between terrorism and weapons of mass destruction: "These same terrorists are searching for weapons of mass destruction, the tools to turn their hatred into holocaust. They can be expected to use chemical, biological and nuclear weapons the moment they are capable of doing so." He did not mention Iraq; but what he meant in that forum was clear when he said: "Some governments"—so not just the Taliban—"still turn a blind eye to the terrorists, hoping the threat will pass them by. They are mistaken. And some governments, while pledging to uphold the principles of the U.N., have cast their lot with the terrorists. They support them and harbor them, and they will find that their welcome guests are parasites that will weaken them, and eventually consume them."[23]

Finally, on November 26, Bush made a short but significant

remark in a press conference. In response to the question, "Does Saddam Hussein have to agree to allow weapons inspectors back into Iraq? Is that an unconditional demand of yours?" the president answered: "Saddam Hussein agreed to allow inspectors in his country. And in order to prove to the world he's not developing weapons of mass destruction, he ought to let the inspectors back in." What followed in the dialog with the journalist is interesting:

Q: And if he doesn't, sir?
A: Yes?
Q: And if he does not do that sir, what will be the consequence? If he does not do that, what will be the consequences?
A: That's up for—he'll find out.

The reporter got his meaning:

Q: Sir, what is your thinking right now about taking the war to Iraq? You suggested that on Wednesday, when you said Afghanistan was just the beginning.
A: I stand by those words. Afghanistan is still just the beginning. If anybody harbors a terrorist, they're a terrorist. If

> they fund a terrorist, they're a terrorist. If they house terrorists, they're terrorists. I mean, I can't make it any more clearly to other nations around the world. If they develop weapons of mass destruction that will be used to terrorize nations, they will be held accountable. And as for Mr. Saddam Hussein, he needs to let inspectors back in his country, to show us that he is not developing weapons of mass destruction.[24]

The Axis of Evil

That was definite enough, and perhaps for precisely that reason the topic once again temporarily disappeared from the president's speeches: after the radio address to the nation on December 22, 2001, there was no further mention of terror. But the silence was deceptive. In his famous "Axis of Evil" speech before Congress on January 29, 2002, Bush asserted the connection between "terrorism," Iraq, and "weapons of mass destruction" as having been proved by recent events, and declared it the linchpin of future policy: "I will not wait on events, while dangers gather. I will not stand by, as peril draws closer and closer. The United States of America will not permit the world's most dangerous regimes to threaten us with the world's most destructive weapons."[25] Today we know that Bush had made up his mind long before. According to an account by Bob Woodward, the president had already decided to invade Iraq on November 21, 2001, that is, just a few days after his combative speech before the U.N.—obviously based on ongoing pre-planned scenarios. His brief remarks to the press of November 26 support these claims (though other observers consider Bush to have been still undecided at the time and instead point to Vice President Dick Cheney and Secretary of Defense Donald Rumsfeld as the driving forces).[26]

At any rate, it was a period when, as we have seen, Bush's semantics discernibly and decidedly toughened. He went from talking about box cutters and airplane attacks to the threat of weapons of mass destruction by terrorists bent on "Holocaust": mediated by the signifier "anthrax," terror became bioterror. And—thanks to the fifty-year-old, familiar signifying string "NBC"—bioterror in turn implied a terror threat of nuclear and chemical weapons. In short: apparently quite naturally, in the space of only four months, the airplane attacks had been transformed into the threat of WMDs and, accordingly, state-sponsored terror. The signifier "anthrax" had begun by taking on new, metaphoric signifiers, but then—and in particular—was metonymically supplemented and finally replaced by other, more comprehensive signifiers. By the end, all that remained was anthrax—huge amounts of anthrax—a deadly instrument in the hands of Saddam Hussein.

The "Axis of Evil" speech of January 29, 2002, announced the new doctrine to the world. It was a thinly veiled declaration of war from a president who could not yet make such a declaration explicitly. Paradoxically, the doctrine was not new, but in the context of 9/11 it acquired a new, dangerous stridency. Bush had obviously been obsessed with the idea of attacking Iraq for a long time. In January 2004, the British *Sunday Herald* reported revelations by Bush's former treasury secretary Paul O'Neill, to wit, "that the president took office in January 2001 fully intending to invade Iraq and desperate to find an excuse for pre-emptive war against Saddam Hussein." According to the paper, O'Neill said that the president was determined to find a reason to go to war. "It was all about finding a way to do it. That was the tone of it," said O'Neill. "The President saying, 'Go find me a way to do this.'"[27]

It was in precisely this frame of mind that Bush reacted to the attacks as early as September 12. Richard A. Clarke reported the following brief exchange with the president: "On September 12th, I left the video conferencing center and there, wandering alone around the situation room, was the president. He looked like he wanted something to do. He grabbed a few of us and closed the door to the conference room. 'Look,' he told us, 'I know you have a lot to do and all, but I want you, as soon as you can, to go back over everything, everything. See if Saddam did this. See if he's linked in any way.' I was once again taken aback," said Clarke, "incredulous, and it showed. 'But, Mr. President, Al Qaeda did this.' 'I know, I know, but—see if Saddam was involved. Just look. I want to know any shred.' 'Absolutely, we will look again.' I was trying to be more respectful, more responsive. 'But you know, we have looked several times for state sponsorship of Al Qaeda and not found any real linkages to Iraq. Iran plays a little, as does Pakistan, and Saudi Arabia, Yemen.' 'Look into Iraq, Saddam,' the president said testily and left us."[28]

Moreover, Cheney, Rumsfeld, and Rumsfeld's deputy Paul Wolfowitz not only worked for years within neoconservative think tanks on an Iraqi campaign but also immediately connected to Iraq in the wake of September 11.[29] According to Clarke, Cheney and Wolfowitz were already focused on Iraq as a terror sponsor at the beginning of their term of office, downplaying Al Qaeda.[30] Rumsfeld is reported to have spoken freely at the National Military Command Center on September 11 around 2:40 p.m. about attacking Iraq. According to a co-worker who made the revelation to CBS a year later, Rumsfeld said that he needed "the best info fast. Judge whether good enough hit S.H. [Saddam Hussein] at same time. Not only UBL [Osama bin Laden]."[31]

The government's hidden agenda was thus pretty clear—there seems to be no reasonable doubt that the government had been considering war with Iraq before 9/11. The only question was how to put together a rationale for the war that would pass public muster—in other words, how to prove the "reality" of the threat to the "world" from Saddam's WMD. The problem for the hawks in the administration immediately after 9/11 was plain: What could Saddam possibly have to do with the airplanes? How could a bridge that did not exist be made to link 9/11 to Saddam?

I would like to argue that the anthrax letters played a crucial role in the search for a solution to this problem. Analyzing the signifiers in Bush's speeches shows that here the letters actually helped. Very quickly and very effectively, the anthrax-derived signifiers "bioterror" and "WMD" encouraged the connection between the terrorism that killed 2,752 people in New York on September 11 and Saddam Hussein's regime.[32] The letters made it possible to leverage public opinion in the service of the neoconservatives' hidden agenda and to make regime change in Iraq appear essential for "world" security. The fact that legal justification was still missing was obvious, as the so-called Downing Street Memorandum has made very clear: Beginning in March 2002, the British government pushed the United States to include the U.N. in the process for purposes of constructing the legal pretext for a war which had already been decided.[33]

In light of the fact that the "Axis of Evil" speech was itself a declaration of war, it was certainly a significant change in foreign policy strategy vis-à-vis the Clinton government. But paradoxically, Bush's policies were not really new. Analysis of the "Axis of Evil" speech and Bush's policies up to that point has made too little of the fact that Bush did little more than repeat essentially word for word key passages from the Defense Act of September 23, 1996. His one (admittedly clever) innovation was to attach the label "Axis of Evil" to some of the states specified in the act.[34] (Then again, even that was not all that new; Ronald Reagan had, after all, christened the Soviet Union the "Evil Empire.") Item 1402 of the Defense Act reads: "Congress makes the following findings . . .

6. As a result of such conditions [fall of the Soviet Union with superfluous NBC weapons capacity and so on], the capability of potentially hostile nations and terrorist groups to acquire nuclear, radiological, biological, and chemical weapons is greater than at any time in history.
7. The President has identified North Korea, Iraq, Iran, and Libya as hostile states which already possess some weapons of mass destruction and are developing others.
8. The acquisition or the development and use of weapons of mass destruction is well within the capability of many extremist and terrorist movements, acting independently or as proxies for foreign states.
9. Foreign states can transfer weapons to or otherwise aid extremist and terrorist movements indirectly and with plausible deniability.[35]

What did Bush really add here? He gave no new twist to the material, apart from letting Libya off the hook in return for its willingness to cooperate following the January 31, 2001, verdict in the Lockerbie trial. Nor did he rethink it in any obvious way. Rather, he asserted with a fair amount of rhetorical flourish connections for which he had not the least shred of empirical evidence. These connections had already been codified as truth, by law, in 1996, and exactly five years later seemed to be confirmed by the September 11 attacks and especially by the anthrax letters. What could be truer than a shocking event like September 11, for which a perfect signifier stood at the ready, eager to embrace it?

Without anthrax, no axis of evil. Yet this axis could not be built on the slim foundation of four airplanes and five letters alone. What is notable about Bush's speech of January 2002 is how the signifiers "bioterror" and especially "weapons of mass destruction" were systematically brought into circulation with reference to September 11 and terrorism. "Anthrax" was first supplemented, then more frequently replaced, by these signifiers, the better to keep the letters in memory. It was precisely this shift that contaminated the consciousness of millions of people and was the crucial element in the slow and very carefully orchestrated refocusing of the public sphere: from the rather unsuccessful struggle of Al Qaeda in Afghanistan to the threat of war against Iraq over the course of the summer and autumn of 2002. For example, on October 7, 2002, Bush characteristically suggested the connection between Iraq and weapons of mass destruction: "The Iraqi dictator must not be permitted to threaten America and the world with horrible poisons and diseases and gases and atomic weapons." That was pure fantasy; but it was what Bush pretended to be protecting America against, and he hastened to add for the umpteenth time that Iraq could supply Al Qaeda with these weapons in order to act without leaving "fingerprints."[36]

October 17, 2002, was an important date in the government's successful new program of speech management. In response to international pressure, the United States introduced a resolution to the U.N. against Iraq omitting any explicit threat of force and accepting a new round of weapons inspection (the famous Resolution 1441). And the Bush administration chose *precisely* this moment to announce that North Korea had for years been pursuing a secret atomic weapons program (obviously not news; and not mentioned in the January 29 "Axis of Evil" speech). In the context of Iraq, the signifier "weapons of mass destruction" had already loosely taken on the connotation "atom bomb." As later revealed, this association was based on a falsified intelligence agency document.[37] Now, the signifier also connoted "North Korean atom bomb," which served to underscore its danger and "truth."

That does not mean, however, that the signifier denotes situations in the real world, as "realistic" epistemologists might desire: that may well be the case, as is easy enough to show for "WMD"—in North Korea, for example—but it can also not be the case, as in Iraq. In any event, it is not the job of the signifier to *reflect* reality. It is rather its function to *try* to describe something by referring to other signifiers. In lending themselves as models, terms, categories, "labels," signifiers reveal perceivable states in the world. Only then do they enable and create reality. But this *attempt* to describe something has a better chance of succeeding when signifiers are associated with, or replaced by, other signifiers that have already proved their "truth."

The next notable event in the series of war preparations occurred on January 31, 2003. At that time the media reported that the controversy over the North Korean atomic program could escalate at any time. The North Korean news agency KCNA was reported as stating that the "military situation on the Korean Peninsula is so tense that a nuclear war may break out any moment."[38] But shortly after 4 p.m. EST, President Bush and Prime Minister Blair appeared briefly before the press and, via CNN, before a world audience.[39] As though it were totally natural, they studiously ignored the threat. Instead, they repeatedly referred to the direct "link" (Blair) between "terrorism" and "weapons of mass destruction"—as a "twin problem" (again, Blair's words; he might more accurately have called it a *twin towers problem*).

Mostly likely, at that very moment, the American preoccupation with the signifier "weapons of mass destruction" in the context of dividing the world into good and evil helped to spur a lone North Korean dictator to play at geopolitics and to talk of nuclear war. But, strictly speaking, the descriptive function of the signifier "weapons of mass destruction" was not what Bush and Blair were after. They only used the North Korea–enhanced *associative* threatening effect of this signifier to suggest the danger of hypothetical Iraqi weapons. This semantic hand waving brings very quickly and somewhat nostalgically to mind the satellite photos of Soviet nuclear missiles in Cuba in 1962. Or the cryptic but obviously well-intended efforts to nail down the type, number, design, and technical details of these same missiles reported in western newspapers up to the 1980s.

In this respect, the Cold War had a kind of truthfulness. With the emergence in 1991 of the "new world order," that truthfulness was replaced by hysterical evocations of the danger of Iraq, an isolated, newly industrializing nation of 22 million inhabitants, controlled by U.N. inspectors and kept in check by continuous satellite and intelligence agency monitoring. But, as already mentioned, these realities were not the issue. At a time when the world was on tenterhooks awaiting concrete proof of an Iraqi threat, Bush and Blair had nothing to say other than what was already a matter of record since September 23, 1996, under Public Law 104–201: that the United States and "the world" risked a direct "threat" of Iraq "transfer[ring]" its "weapons of mass destruction" to "terrorist groups." Now, inevitably, the signifier "weapons of mass destruction," codified in 1996, began to create its own reality.

And in fact, as is well known, the rhetoric was effective. In January 2003, surveys showed that 50 percent of Americans were convinced that Saddam Hussein was responsible for the attack on the World Trade Center and was threatening the United States with unspecified "weapons of mass destruction."[40] In the face of an ever-growing likelihood of war against Iraq, American fear of "bioterror" also increased. Like a self-fulfilling prophecy, this fear served in turn to justify a "preventive" war. Where was there room for reasonable doubt? In his crucial speech of February 5, 2003, before the U.N. Security Council, Secretary of State Colin Powell produced such evidence as he had—snippets of tape recordings and satellite pictures—that convinced most, though not all, of the members of the Security Council. Powell looked very grave as he put his credibility on the line: "My colleagues, every statement I make today is backed up by sources, solid sources. These are not assertions. What we're giving you are facts and conclusions based on solid intelligence."[41]

Let us consider just the critical statements regarding bioweapons. Powell showed the U.N. delegates a small container of white powder—"a teaspoon of dry anthrax"—to demonstrate that Iraq had produced 25,000 liters of the agent and had loaded much of it into 400 bombs. ("This is evidence, not conjecture. This is true.") Powell further showed slides of mobile research laboratories and mobile production facilities for bioweapons, of which Iraq possessed at least seven units, each containing two to three trucks. "Ladies and gentlemen, these are sophisticated facilities. For example, they can produce anthrax and botulinum toxin. In fact, they can produce enough dry biological agent in a single month to kill thousands upon thousands of people." Then Powell showed a video of an Iraqi F-1 Mirage jet aircraft that could spray 2,000 liters of anthrax from its modified fuel tanks. In addition, Iraq planned to build UAVs (unmanned aerial vehicles), remote-controlled drones for delivering biological agents, "an ideal method for launching a terrorist attack using biological weapons." In other words, "There can be no doubt that Saddam Hussein has biological weapons and the capability to rapidly produce more, many more. And he has the ability to dispense these lethal poisons and diseases in ways that can cause massive death and destruction."[42]

All Wrong

We know now that these weapons did not exist—at least not between 1995 and March 2003.[43] In December 2003, writing in the *New York Review of Books,* Thomas Powers took the trouble to recite all twenty-nine (by his count) of the hard facts presented by Colin Powell as evidence of Iraq's arms program. He compared them with the findings of David Kay, whose commission searched the defeated country for the weapons in question. The result of this comparison was sobering in its utter clarity: twenty-nine "not found's."[44] We also know that each further formulation of the alleged threat was based on dossiers falsified or inflated by Pentagon hawks over the repeated objections and risks to their reputations of members of the intelligence agencies. In the end, CIA director George Tenet accepted full responsibility, as is his office's practice.

In an interview with *Vanity Fair* at the end of May 2003, then U.S. Deputy Secretary of Defense Wolfowitz confessed that the "weapons of mass destruction" rationale for the Baghdad campaign was a media strategy chosen "for bureaucratic reasons," meaning the need to obtain consent within the administration, most likely the conflict between the State Department and the Pentagon: "We settled on one issue, weapons of mass destruction, because it was the one reason everyone could agree on."[45] In 2004, the Carnegie Endowment for International Peace determined that the weapons did not exist or existed no longer.[46] And in its report to the president of March 31, 2005, the Commission on the Intelligence Capabilities of the United States Regarding Weapons of Mass Destruction concluded flatly, "These assessments were all wrong."[47] Even the CIA could not get around the statement that—for example, concerning the alleged bioweapons program—it was not possible "to establish the scope and nature of the work at these laboratories or determine whether any of the work was related to military development of BW agents." Nor that "in spite of exhaustive investigation, ISG [Iraq Survey Group] found no evidence that Iraq possessed, or was developing BW agent production systems mounted on road vehicles or railway wagons."[48]

But the staging had been perfect, and it was based on a structural gap: the complicity in the anthrax letters could not be established. The Iraqi weapons of mass destruction had to continue existing as a vague threat and a dangerous possibility even after the destruction, confirmed by the U.N. inspectors, of all production capacities and weapons systems between 1995 and 1998. Because just as the hysteric's desire is sustained only so long as it remains unsatisfied, the evidence that the Iraqi weapons of mass destruction either did not exist or had been destroyed was not acknowledged or was ignored.[49] U.N. weapons inspector and former Marine officer Scott Ritter reports that in 1998 Dick Spertzel, a bioweapons expert for the U.N. team, refused to investigate the alleged anthrax stocks: "The Iraqis repeatedly asked him to bring sophisticated sensing equipment to test for biological weapons. He consistently said he wasn't going to carry out investigations that provide circumstantial evidence to support Iraq's contention they don't have these weapons."[50]

In other words, recalling the nature of hysterical desire, it follows that only the perpetual maintenance of "suspicion" enables what Emmanuel Todd calls "theatrical micromilitarism": the overwrought performance of a military power designed to hold the entire world in suspense.[51] As early as April 20, 2003, the *New York Times* reported that the United States was planning a "long-term military relationship with the emerging government of Iraq, one that would grant the Pentagon access to military bases and project American influence into the heart of the unsettled region."[52] Indeed, in his interview with *Vanity Fair,* Wolfowitz stated that the principal effect of the war in Iraq and the victory over Baghdad was to allow the United States to withdraw its troops from Saudi Arabia, where they were felt to be a substantial provocation. America could then construct a new, stable military presence in the neighboring country and thus the center of the Arab world. That today things in Iraq look anything but stable and that the United States would like to pull its troops out of the self-created inferno sooner rather than later—assuming it can—in no way alters this original intention.

And it had long been obvious. In a "Parody" (that could as well have been captioned "Freudian Slip—A Really Big One") the July 7, 2003, issue of the neoconservative, pro–Iraqi war *Weekly Standard* published a humorous take-off on recent history. A full-page "ad" carries an invitation to a cruise titled JOIN US AS WE CONQUER THE WORLD! that is plastered over a photo of an aircraft carrier. The ad trumpets "Eight imperial days & seven unilateral nights aboard the USS *Benevolent Hegemon*" with Dick Cheney, Richard Perle, Ariel Sharon, Elliott Abrams, Donald Rumsfeld, and Paul Wolfowitz as guides and evening entertainers.[53] Attractions include "Neocon Hot Bikini Contests," and such games as "Help Us 'Find'[!] Weapons of Mass Destruction in Iraq" and "Name That Evildoer: The Dead Terrorist Nostalgia Game." There is even a "Special Seminar: Reading between the Lines of George W. Bush's Major Speeches." The trip includes a flight to Jerusalem, and ports of call at "Greater Israel (Hebron, Amman, Damascus)," "Liberated Iraq," and "Iran." Travel costs are not stipulated. The ad ends with the exhortation: "Don't Consult with Your Allies! Sign Up Today!"

3

The Cobra Event

New York, end of the 1990s. In the loft of a renovated warehouse on the west side of Union Square, seventeen-year-old Kate Moran was dressing for a chilly day in April. She was the only child of a very busy investment banker and a lawyer, both of whom had already left for work, and the housekeeper was urging the curly redhead to hurry. Kate's nose was running and she wasn't feeling that great. But she got on the subway to go to school. Once there, her vague malaise turned to real sickness. Her mouth was blistered and burning, and clear mucus poured out her nose. She started shaking and had to lie down; she was having trouble forming words. Then her body tensed, her legs lashed out, and she went into epileptic seizures.

"Her teeth sank into her lower lip, cutting through the lip, and a run of blood went down her chin and neck. She bit her lip again, hard, with ferocity, and she made a groaning animal sound. This time, the lip detached and hung down. She pulled her lip in, sucked it into her mouth, and swallowed. Now she was chewing again. Eating the inside of her mouth, chewing her lips, the inside of her cheeks. The movement of her teeth was insectile, like the feeding movements of an insect larva chewing on its food: intense, greedy, automatic—a kind of repetitive yanking at the tissues of her mouth. Her tongue suddenly protruded. It was coated with blood and bits of bloody skin. She was eating her mouth from the inside."[1] Blood streamed from Kate's nose. Her spine bent until it cracked. She died after hours of indescribable agony.

Richard Preston's novel *The Cobra Event,* published in 1997, begins quite alarmingly, and readers are warned that what follows will be more of the same. We learn that Kate was the first victim of an unknown virus released right in the middle of New York. Soon more bodies are found, showing the same frightening symptoms, especially signs of self-destruction: autopsies show that all the victims of the puzzling new illness began to devour themselves before their brain turned into a gelatinous mass and put an end to the biting and the seizures. The authorities react very quickly. First, Alice Austen, a young doctor with the Centers for Disease Control and Prevention (CDC) in Atlanta and the protagonist of the novel, is sent to New York, where she searches for an epidemiological pattern underlying the scattered medical cases. It becomes apparent that the unknown source of the illness has to be a pathogenic microorganism, and that it is very unlikely to be due to natural causes.

The FBI and the National Security Agency are promptly called in, because the growing number of dead and the unexplained symptoms suggest a terrorist strike. At the same time, we find out that although the perpetrator worked alone, he must have had some connection to production of bioweapons in Iraq. Finally, a massive police and intelligence agency operation succeeds just in time to prevent the madman, named Archimedes, from detonating two virus-charged bombs in the New York subway. Archimedes himself is already infected with a viral agent grown in his unprepossessing apartment, and he too dies an autocannibalistic death. The story concludes with two dozen dead and the illusion that large-scale decontamination of the crime scene has killed all the virus. But on the last page of the novel, the virus reappears: it has found a new host in the rats inhabiting the city's underground.

A nice thriller to pick up at the airport bookstore, obviously well investigated and not badly written, with, as a finale, a conventional but suspenseful chase through the dark tunnels of the New York subway. Just a novel? In the context of "bioterror," *The Cobra Event* cannot be underestimated: the book was a signifier bomb. The "bioterror" virus, which up to then had existed in natural symbiosis with the specialist milieu of the intelligence agencies, was transformed through Preston's carefully carried out mutations into a highly contagious species. As early as 1998, the virus was able to jump the species barrier and infect ordinary readers like the president of the United States. And Bill Clinton was a superspreader.

A Hybrid Virus

Before going into this history, it is worth taking a closer look at the novel and its author. Richard Preston obtained a doctorate in English from Princeton and won fame with the publication of *Hot Zone,* a nonfiction account of an Ebola virus outbreak near Washington, D.C. In the Acknowledgments to *The Cobra Event,* he stresses the thriller's semblance to reality—which scarcely clarifies the relationship between fiction and reality but rather clouds it further. *The Cobra Event* ostensibly reflects the position of science, geostrategic reality, and intelligence down to the last detail, an impression the author does not try to dispel. In a very effective way that underscores the unverifiable truth of this work of fiction, Preston acknowledges his informants at the FBI, the intelligence agencies, and the Pentagon. These sources provided him not only with insider information but also, for instance in the case of the Nobel Prize winner and Clinton advisor Joshua Lederberg, with tips on how to construct the story. Lederberg is said to have told Preston to let the assassin die of his virus at the end, to deter possible real-life imitators.[2] *The Cobra Event* was deliberately written to have an effect.

The know-how Preston displayed in the novel earned him a reputation as an expert on biological warfare. According to an interview in the online magazine *The Thresher,* he stated, not without pride, that the U.S. National Security Agency was supposed to have asked with a certain amount of excitement where "all this classified information" was coming from.[3] In any event, not only did the president take Preston seriously; in

1998 the author was also invited to appear as an expert before the Senate Subcommittee on Technology, Terrorism and Government Information, which was holding a joint meeting with the Select Committee on Intelligence.

The novel achieves the reality effect to which Preston owes his reputation by fairly simply but effectively recombining various settings and action into a new, hybrid narrative. First, the obvious setting of New York, "end of the 1990s"; second, the narrative device of the probably more or less "true" but "invisible" (Preston) history of biowarfare; third, a mix of indistinguishable half-fictional, half-real elements that support the story. Thus, following the description of Kate Moran's death the text proceeds to depict the then still secret, or less well known but allegedly very successful, American series of tests with biological warfare agents in the second half of the 1960s. The story then introduces the reader to the young (fictitious) epidemiologist Alice Austen in the real CDC, preparing for her flight to New York, and finally jumps to the U.N. weapons inspectors in the Iraqi desert. Later chapters detail bioweapons production in the former Soviet Union and correspond to some extent with authoritative accounts by Miller, Engelberg, and Broad.[4]

Indeed, everything centers on Iraq. The title of the first

chapter, "West of Babylon," is an ominous reference to ancient history. It describes a convoy of white four-wheel-drive vehicles carrying U.N. weapons inspectors (the "United Nations Special Commission Biological Weapons Inspection Team Number 247—UNSCOM—it was called") rolling through the desert, followed by Iraqi officials "in a rattletrap column of vehicles" with cute geopolitical connotations: "beat-up Toyota pickup trucks, smoking dysfunctional Renaults, hubcapless Chevrolets, and a black Mercedes-Benz sedan with tinted windows and shiny mag wheels."[5] Suddenly, retired U.S. Navy commander Mark Littleberry commands his colleague, FBI bioweapons expert William Hopkins, to steer their Nissan Pathfinder 4×4 off course, breaking away from the convoy and heading full-speed toward a building on the horizon—a "snap inspection," as Littleberry informs his astounded U.N. colleagues and the leery Iraqis by radio. The commander of the U.N. unit, Frenchman Pascal Arriet, "a crackly voice" over the radio "growing more and more hysterical," tries in vain to stop the Americans. Littleberry and Hopkins's target is a factory that they suspect contains fermenting tanks for making germ warfare agents and bioreactors for breeding viruses.

In the Al Ghar Agricultural Facility they meet Dr. Mariana

Vestof, clearly modeled on the notorious Dr. Rehab Taha, but again a hybrid version. Vestof is introduced as a Russian living in Geneva and working there at the biotech company BioArk, which has also built the facility in Iraq. The two American U.N. inspectors now proceed further into the building, through a jungle of stainless steel tanks and pipes, followed by a very indignant Dr. Vestof and the Iraqi technicians. Littleberry sees a door and ducks into it. He enters a Level 3 laboratory, then an entry chamber that leads into a Level 4 zone where "two people wearing biohazard space suits" work. A diesel motor starts up. "A crack of gray desert sky opened up over Littleberry's head. It widened. The hot lab was inside a truck. It was a mobile hot zone, and it was beginning to pull away from the building."

Littleberry's FBI colleague Hopkins manages to jump into the still open rear door of the truck and to take a sample with a medical swab stick before a man swears at him in Russian and throws him out again. "The doors of the truck slammed shut, and it roared off down the road." The truck vanished—and floated around various classified files until the alleged mobile bioweapons laboratory turned up in Colin Powell's presentation on satellite photos in the U.N. Security Council on February 5, 2002. After the war, the mobile lab was allegedly "found" many times, only transformed on closer inspection back into a fire truck or some similar vehicle.

Now the novel begins to mix story lines, or rather lines of reasoning. Preston's intent is to show that the future bioterror, that is, biowarfare, will be based not on naturally occurring disease agents but on designer weapons, against which there is no defense. The Cobra virus is such a weapon. In the course of investigations in New York, in which the insubordinate American (now former) U.N. inspectors also take part as experts, we learn that Cobra is similar to a nuclear polyhedrosis virus (NPV) that kills butterflies but is harmless to humans. In the nerve cells and brain of the insect the virus multiplies incredibly quickly, until it totally invades the animal's body. In the style of good popular science, Preston includes a photo of the virus magnified 25,000 times as well as bits of genetic sequences, to bolster the idea that "[everything] is real" (the same goes for the glossary of biological terms annexed at the end).[6]

For the NPV to be lethal to humans, biochemists from the former Soviet Union, from Iraq, and from the American subsidiary of BioArk in Geneva (Preston strives, somewhat hypocritically, for geopolitical balance) have to incorporate three additional sequences into the viral genome. The now utterly deadly hybrid contains, as a starter, the common cold virus (to speed transmission of the virus from the mucus membranes of the nose to the brain), next, the gene for smallpox—"that's ancient and old and smells like Russia," as Littleberry says—and finally, the gene from Lesch-Nyhan syndrome.[7] Lesch-Nyhan is a degenerative brain disease caused by an extremely rare gene defect that results in spastic movements of the body and self-mutilation such as biting of lips and fingers.[8] The hybrid virus, which penetrates the brain by way of the neural pathways, ultimately unleashes "brainpox": as victims begin to devour themselves, old Russia literally pounds their brain into mush. The auto-cannibalism that is manifest in extraordinarily aggressive behavior toward others turns victims into wild beasts for the final hours of their lives.

A touch wary of plainly implicating Iraq, Preston peoples his story with craven Swiss racketeers in cahoots with smart young geneticists in American-based Swiss-owned startups. But the novel's guiding impulse is the suggestion of a lethal Russia-Iraq connection. The *Thresher* interviewer grasped the message in telling Preston: "I've been thinking that we're reaching the point where we could generate a Doomsday Machine—and one that's lowtech and easily accessible. Mesopotamia, for example, could whomp up some high-lethal high-infectious bugs and then hold its neighbors—or the world—hostage. I mean, if I were a small fanatical country with a last-ditch agenda, this is the way I would go." Preston doesn't contradict him; on the contrary: "Yes, exactly"—meaning Iraq, a small "fanatical" country capable of taking the world hostage with inexpensively assembled bioweapons.[9]

Reading the book today makes your jaw drop. In the years since it was written, reality appears to have more than caught up with fiction—and, since the fall of Baghdad in April 2003, "reality" has turned back into fiction. The suspicion was opportune, and even in the novel itself sparked fear of war: "The possibility that the deaths in New York were a terrorist event being sponsored by Iraq weighed on him [Hopkins]. He discussed it by phone with Frank Masaccio [the FBI officer coordinating the investigation]. Masaccio was very disturbed by this. 'If this is terrorism sponsored by a foreign government, Will, this could start a war.' 'I know, Frank,' Hopkins said."[10] A statement at the end of the novel cites the "existence" of "evidence . . . of a continuing biological-weapons program in Iraq, a program that had apparently moved into the genetic engineering of viruses."[11]

Preston's biological descriptions seem totally convincing. Today, DNA recombination is a routine activity for any decently trained molecular biologist anywhere in the world. The first successful experiments with recombinant DNA carried out at Stanford University in the early 1970s ignited a wave of self-questioning among microbiologists as well as sharp-witted military experts about the possible consequences—military, epidemiological, and medical—of unnatural hybrids produced with the new technology. At the so-called Asilomar Conference of 1975, the worldwide scientific community of molecular biologists and geneticists agreed to strict safety standards for genetic experiments and to respect the rules and limits of the Biological and Toxin Weapons Convention of 1972. The whole problem of advanced, that is, genetically engineered, bioweapons turns on whether this self-restraint can be universally observed or enforced, as the case may be. Nervousness about biological weapons programs is rooted in the fact that too many states appear unwilling to adhere to the convention. Yet evidence of weapons involving genetically modified bacteria or viruses could not actually be produced even for the Soviet Union, with its admittedly enormous bioweapons program. To date, all we know for sure is that genetic engineering carries the potential for malicious or deadly abuse.

Preston constructed his novel such that the reader has no

way of distinguishing between genuine biotechnological possibilities and pure fiction; rather, he systematically blurs this border. And that's not all: The portrayal of genetic engineering in *The Cobra Event* bears only superficial resemblance to good popular science and is actually an insidious form of disinformation. Lesch-Nyhan syndrome is not caused by a gene sequence that can be arbitrarily inserted into or deleted from a string of DNA, but by a defective *(missing)* gene on the X chromosome in *every single* cell in the body. Since missing genes obviously cannot be recombined, and moreover because viruses only ever attack a very small number of cells, it is impossible for such a genetic defect to infiltrate the body à la *Cobra*.

Hansruedi Bueler, a research fellow at the Institute of Molecular Biology at the University of Zurich, works on mechanisms of brain degeneration and related gene therapy. He explained to me the fundamental problem of the gene transfer that Preston fantasized in 1997: "To unleash a genetic illness, in this case, to cause brain disease, the gene would have to be present and active in a majority of brain cells. To date, not a single one of the viral vectors [a harmless virus used to carry gene sequences] can do that. It's an ongoing struggle for gene therapists." Only by injecting a viral vector "directly into the brain tissue with a fine needle do we obtain gene expression in a few thousand to tens of thousands of cells in the treated region. So I have to wonder how this butterfly virus could be 'applied' to infect brain cells, that is, how you could physically manage to do that."[12] Walter Schaffner, director of the same institute, adds: "An insect virus with such a hodgepodge of foreign genes would most probably be quite harmless both to butterflies and to people."[13]

Preston's superbug is not at all plausible in a genetic engineering sense. Yet he was very successful in making the dangers of his fictitious hybrid the epitome of a bioterroristic threat. He was particularly serious about the smallpox virus spliced into the DNA of the NPV. But the difference between this fictional hybrid and the genetically modified bacteria or virus—that is, designed to resist known antibiotics—is lost on the public. In an article published in the *New York Times* on November 7, 1997, Preston maintained that, "based on reports I've heard from members of United Nations inspection teams, top government officials and scientists, as well as photographs and United Nations documents," Iraq possessed genetically modified cultures of anthrax, plague, Ebola, botulin toxin, and smallpox. "Intelligence experts . . . believe," Preston continued, that even deadlier versions of these genetically modified organisms were circulating on the international black market. What would happen if they were used against defenseless American cities? The result would be "terror in slow motion, an unrolling horror with a death toll equivalent to dozens of Oklahoma City bombings day after day."[14] Compared with this article, which caused a sensation, the death toll in *The Cobra Event* seems relatively small; on the other hand, the Cobra virus was so realistic that neither *Cobra Event* readers nor readers of the *New York Times* could tell fiction from reality.

In this way, the threat scenario could be developed and recombined, as it were, with earlier examples from the Cold War to breed new, hybrid varieties of fear. In April 1998, when Preston appeared before the joint Senate committee, he cited "100 to 400" strategic warheads on Soviet intercontinental missiles filled with smallpox and plague supposedly targeted to American cities. He continued: "I have no idea where those biowarheads are now . . . Russian military people have never said these warheads were destroyed." And? "One can wonder if other countries, such as Iran or Iraq, have obtained examples of the biowarheads, for use as study models for their own missile programs."

Warheads bearing genetically engineered invisible killers:

that is the suspicion on which the novel is based, and Preston could only reaffirm it to the Senate. Instead of proof, he served up an abbreviated lesson in popular science. In a manner one reserves for the slow-witted, he patiently explained that weapons-grade particles are so fine that they disperse easily in the air and can fly for miles: "I'll give you a demonstration using harmless baby powder," he said. "This illustrates what a bioweapon really looks like in the air—it disperses and becomes invisible and undetectable."[15] Baby powder on Capitol Hill! The various powders in the "anthrax" letters following September 11 had a surprisingly ubiquitous precursor.

Powerful words, fine powder, and the usual suspects—but could Preston, the new star of bioweapons experts, actually be right? Jonathan B. Tucker, former U.N. weapons inspector in Iraq and director of the Chemical and Biological Weapons Nonproliferation Project at the Monterey Institute of International Studies, is one of the experts Preston supposedly relied on. As such, he was hardly ill-informed. When asked in an interview with the online magazine *Salon* in 1997, "Do we know for sure that Iraq has these genetically engineered toxins that Preston describes?" Tucker answered, "No. I read

Preston's article, and it knocked my socks off." But: "He doesn't say who his sources are; I'd love to know who they are."

The same question applies to the black market, to the joint undertakings between Russians and Iraqis, and to the assertion that Iraqis had bought smallpox virus on the black market: "It's shocking, if true," but by the way, smallpox is not a very good bioweapon because it is highly contagious and therefore uncontrollable. Tucker inferred that Preston was presenting speculations as facts: "These are very sensational allegations, and it's irresponsible for him to make them unless he has really solid evidence for them." At any rate, the U.N. weapons inspector had found no proof that Iraqi weapons were equipped with genetically modified biological agents.[16] Tucker's opinion is echoed by Martin Schütz, a microbiologist active in Iraq in 1995 as an UNSCOM inspector in the context of the twenty-fourth bioweapons mission, and currently director of the biology department of the Swiss defense department. Asked about Preston's contention that bioproduction equipment could be fitted onto trucks, Schütz answered: "Iraq never had a mobile Level 4 containment laboratory. Indeed, to date no one has definitively been able to discover a single mobile production unit for B-weapons, although U.S. intelligence in particular has always insisted that they exist."[17]

To support his thesis of work on advanced, that is, genetically modified weapons in Iraq, Preston maintained in his article in the *New York Times* that "American inspectors on the United Nations teams have been using a simple tool to try to ferret out bioweaponry in Iraq: a medical swab. It is used to collect dirt from the corners of buildings, liquids dripping out of bio-reactors, crud from unwashed test tubes, soil. Using such samples, machines can find the molecular fingerprint of a bioweapon. The Iraqi 'minders,' or escorts, 'kind of clump up in a group and get real agitated whenever we whip out our swabs,' one American inspector told me."[18] That is, nobody but the American weapons inspectors went to the trouble of using medical swabs to test laboratories, production plants, and weapons stockpiles for genetically modified bacteria such as anthrax or viruses, and the Iraqis were accordingly uneasy. These are the same swabs that Hopkins and Littleberry use in the novel to catch out the Iraqis—"He hurled himself toward the hot zone, a swab stick held in front of him."[19] In Preston's 1997 editorial in the *New York Times,* the medical swabs are presented as cast-iron evidence, just like the baby powder in the Senate. The story of Americans hell bent on finding out and Iraqis who "kind of clump up in a group and get real agitated" when the fearless U.S. inspector pulls out his swab stick was too good not to be true.

Except that the issue wasn't the swab sticks. Of course, as Martin Schütz acknowledges, UNSCOM inspectors (and not only American!) sampled bombs, warheads, and the respective surroundings (not exclusively but also using swabs) in order to verify the declared bioweapons. But in so doing, the inspectors encountered a problem that had not previously emerged in any discussions on Iraqi bioweapons. "The interpretation of the results [of these samples] was difficult, because anthrax is endemic in Iraq and it wasn't possible to differentiate the endemic anthrax strains from those used in the bioweapons program."[20]

How could we have forgotten? Ordinary anthrax is carried by donkeys and camels back and forth across the desert, and in its wild form can also fly hither and yon, into every crack and over every fence. It clings to goatskins and, according to WHO, regularly infects humans—in this case, Iraqis.[21] This ordinary anthrax gets mixed up with western "anthrax"—weaponized, baby powder, genetically engineered, whatever—at any rate, the very same "anthrax" that the American inspectors are certain exists. But whenever they take out their swabs, a goat gets in the way. It's a real dilemma: Anyone waving a swab stick in the Iraqi desert is bound to pick up ordinary anthrax. On the other hand, anyone using a medical swab as a fancy secret weapon—like the hero in Preston's novel and the sources he cites in the *New York Times*—cannot prove that he has found anything. So the swab stick itself becomes the evidence. And isn't that the point of the novel? To show that something is always sticking to a swab, even if it is just a suspicion?

The President

In 1998 the world learned ad nauseum and in trivial detail that in his second term, a lonely Bill Clinton would occasionally while away the hours in the White House with Monica Lewinsky. What is less advertised is that the president also read late into the night. In a January 21, 1999, interview with Judith Miller and William Broad of the *New York Times,* the first he granted since news of the Lewinsky affair broke, Clinton spoke of the terrorism threatening America and about the books he was reading: "I've had all kinds of—I also find that reading novels, futuristic novels—sometimes people with an imagination are not wrong."[22] Clinton really read quite a lot. Richard A. Clarke recalled with astonishment his "eclectic" choices and his habit of devouring a book a night—since the 1995 sarin subway attack in Tokyo by the Aum Shinrikyo sect, especially "fictional accounts."

Clarke mentions, for example, *Rainbow Six* by Tom Clancy.[23] In this 900-page thriller, which came out in 1998, the owner and managers of a large American pharmaceutical company (who also happen to be radical environmental "believers") are preparing to disseminate a secret hybrid virus to free the planet of the parasite *Homo sapiens.* Clancy endeavors to portray the ideology of the characters as a copy of Soviet totalitarianism and even includes an Iranian bioweapons attack that killed five thousand people in the United States.[24] But the plot, though exciting, is politically very abstruse and far from any reality. No wonder that Clinton was far more impressed with another, somewhat earlier book. Responding to a leading question from the *New York Times* journalists, he said: "Preston's novel about biological warfare, which is very much based on—

Q: *Hot Zone* or *Cobra Event?* Which one impressed you?
A: *The Cobra Event.*

Q: That's the one.

A: Well, *The Hot Zone* was interesting to me because of the ebola thing, because that was a fact book. But I thought *The Cobra Event* was interesting, especially when he said what his sources were, which seemed fairly credible to me.[25]

Just like everyone else, Clinton was totally surprised by the apparently reliable sources on which the thriller's author based his book. For his part, Preston reported how Clinton had called his political adversary and old buddy, southerner Newt Gingrich, and said, "You have to read *Cobra Event,* and then get ahold of Richard Preston. Find out what he knows and what we need to do," and how that led to detailed discussions with Gingrich.[26] It does seem a little peculiar that in dealing with this "strategic" question, both the president and the Speaker of the House had to fall back on the research of a thriller author to obtain—alleged? real?—information, not to mention that the U.S. CIA, the FBI, and the Pentagon preferred to confide their information to a novelist rather than to the president and commander-in-chief. But it is no secret that it happened that way. Since 1993 Clinton's own secretary of the Navy, Richard Danzig, had been contemplating a possible bioweapons threat to the American armed forces. Danzig was said to be sharply critical of the lack of coordination among the different, generally low-level experts in the various departments of the Pentagon and the intelligence agencies. But his criticisms were not well received.

Danzig met Preston at a lunch in 1996. Fresh from his success with the *Hot Zone,* Preston was doing research for a new novel about an anthrax bioterror attack on New York, and Danzig soon began systematically putting him in contact with informants at the Pentagon and the FBI. It was an FBI source who argued Preston out of the "too real" anthrax idea and recommended that he construct a fictional hybrid virus instead.[27] With obvious success: So impressed was Clinton not just with Preston's sources but also his fiction that he appropriated the latter as his vision of the future. This vision was similar to the strategic threat scenarios of some of the Pentagon planners, which is to say that it convinced neither the largely skeptical military nor Clinton's advisors. The president, however, was aghast on reading the book (which begins with the awful death of Kate Moran, a russet-haired only child like his then seventeen-year-old daughter Chelsea)—"petrified," as the *New York Times* reported and as is trumpeted on the book's back cover.

Clinton promptly internalized *Cobra Event*'s plot: "It is a near certainty that at some time in the future there will be some group, probably a terrorist group, that attempts to bring to bear either the use or the threat of a chemical or biological operation, I would say that it is very likely to happen sometime in the next few years." Later in the interview, the president specified: "Let's suppose some terrorist hired a genius scientist and a laboratory to take basic anthrax and put some variant in it that would be resistant to all known anthrax antidotes." But who would plan such a thing?

A: We know Osama bin Laden's network has made an effort to get chemical weapons.
Q: Biological or just chemical?
A: Well, we know they've made an effort to get chemical weapons; they may have made an effort to get biological weapons. We do not know that they have them.[28]

A quick transition from chemical to biological weapons, and the suspicion was planted. Richard Preston's "black market" thesis (according to which groups such as Osama bin Laden's are able to obtain advanced biological agents, and in which Russia and Iraq play a central role) was something Clinton had already assumed without proof. We do not know—but it could be.

One reason for this newly gained conviction lay in Clinton's then budding interest in gene technology. At the end of December 1997, Clinton was thoroughly briefed by the geneticist Craig Venter about the possible relationship between genetic engineering and bioterrorism. Not coincidentally, the discussion included smallpox virus. Venter himself had shortly before been on the board of the Institute for Genomic Research and had worked on mapping the smallpox virus genome. Now, he warned Clinton, such research could be used to reconstruct smallpox from other genetic sequences or through recombination to make existing smallpox even more dangerous. In short, to build a superbug. What that means was detailed in a book just published that Venter "urged" Clinton to read.[29]

Following his remarks concerning Osama bin Laden in the January 1999 interview, Clinton continued: Even if it all sounds a bit like science fiction, the logic of the Human Genome Project implies that advanced genetic alterations become possible—and that has implications for biological attack weapons as well as biological defense agents. *The Cobra Event* elaborates this line of argument: Preston has the inspectors use a small handheld device that they carry in a Halliburton suitcase (Halliburton doesn't only supply equipment for oil fields; it is also a big provider of technology for dismantling atomic and biological weapons production plants and associated warheads).[30] The device, a biosensor, can analyze in seconds the genetic sequence of any bacteria or virus captured on a swab and can identify the sequence by comparing it with a data bank. In the *New York Times* interview, these are the "machines [that] can find the molecular fingerprint of a bioweapon."

The idea fascinated Clinton: "We may have to depend upon the Genome Project, interestingly enough, because once the human genes' secrets are unlocked, then—if you and I think we've been infected, they could take a blood sample, and there would be a computer program which would show us if we had—let's say we had a variant of anthrax." Scientists in rogue states with some knowledge of genetics could probably manufacture hybrid bioweapons. But what the Human Genome Project promised in decoding the human genome was to develop custom-made vaccines against each new, genetically engineered bacterium or virus and thus to ensure quick and efficient protection from any bioweapon: "What you would want is to be able to take a blood sample, do an analysis, put it through a software program that had already been developed, and say, okay, here is—this is how the genes are different, this is the difference. And then, presumably, not too long after we've developed this, they will already know, well, therefore, this is how you should—how you should change the vaccine . . . I know this is kind of bewildering, but keep in mind this is actually good news because, if there were no Genome Project, if there were no rapid way to do quick analysis that would go right to the tiniest variant, we would be in trouble."[31]

Clinton believed that there were critical phases in history where great shifts in power occurred because offensive weapons achieved a technological lead over defensive weapons, for instance, in the decline of bronze weapons with the advent of iron ones, the victory of gunpowder over armor, or—according to Joshua Lederberg—the fast cavalry that "enabled Asians to sweep through Europe."[32] Something of this fantasized Asiatic danger threatens anew, if rogue states and terrorists manage to get access to genetically altered pathogens such as smallpox virus; the Human Genome Project assumes a vital importance.

Acting Out

The political effects of Clinton's reading were evident—and influential. From the beginning of 1998, the president addressed bioterrorism in every one of his speeches, calling it the "drawback" of globalization, and to a degree that his advisors and those around him found obsessive.[33] In his State of the Union speech of January 2000, Clinton gave the example that in the near future the security of the United States would be threatened by "narco-traffickers and the terrorists and the organized criminals . . . working together, with increasing access to ever more sophisticated chemical and biological weapons."[34] As early as January 1998, Clinton was demanding that his advisors and high-level Pentagon officials read *The Cobra Event,* and in March the White House conducted a secret exercise that involved playing out a terrorist attack with hybrid smallpox virus—*The White House proudly presents:* Cobra *live . . .*[35]

On April 10, 1998, in the Truman Room, Clinton together with Defense Secretary William Cohen, Attorney General Janet Reno, Health Secretary Donna Shalala, CIA Director Richard Tenet, and National Security Advisor Sandy Berger received seven of the leading American experts in biological weapons systems and public health, including Lederberg, Venter, and Barbara Hatch Rosenberg, an expert on biological weapons with the Federation of American Scientists.[36] Clinton also wanted to talk to these specialists about Preston's book, which some of them apparently found embarrassing. On the other hand, they knew that the meeting with the president presented a unique opportunity. For Venter, *The Cobra Event* and now also this meeting meant a chance to press Clinton to increase government funds for gene sequencing and to plug his biotech company Celera Genomics, then already in the works. Another expert, Thomas Monath, lobbied successfully to stockpile vaccines against anthrax, smallpox, and similar biological agents, and shortly thereafter was accused in the press of having used the meeting to further the commercial interests of his troubled vaccine company OraVax.[37]

The health experts tried at length to make Clinton see that the United States' disastrously neglected health service urgently required additional resources both to handle routine infectious diseases as well as, in extremis, a bioterror attack. For public health sector officials, this alleged "bioterror" threat presented a real opportunity to obtain badly needed resources. According to Miller, Engelberg, and Broad, regarding the situation in 1998, "Almost half of all local health departments did not have the use of E-mail; at least one thousand of them had no access to any on-line or Internet service; . . . 70 percent of health directors had little or no expertise in using computer services. Some local health departments could not even afford a microscope, much less sophisticated computer technology."[38] It is not hard to see the lobbyists' point. The requested budget increase for 1998 included $10 billion for bioterror alone—above and beyond existing expenditures for antiterror efforts of the various U.S. agencies that at the time amounted to $11 billion. They wanted to see as much as possible go into the public health service.

However, these efforts at public health service reform met with little success. When, on May 22, 1998, Clinton submitted his ambitious "germ defense" plan, its focus was on stockpiling vaccines and defensive military measures. Only the military measures were spared a volley of criticism from all sides. The National Guard, for instance, was equipped with a range of mobile high-tech biological-agent rapid-detection labs. But when, in March 1998, the Pentagon began vaccinating troops in the Middle East against anthrax, these measures, too, were denounced.[39]

But that wasn't all. Instead of addressing the obvious deficiencies in the public health service, the government occasionally simulated bioterror attacks in a sort of collective acting-out process. One of the first such major exercises in Washington involved a lone perpetrator in his cellar cultivating anthrax bacteria that an extremist group then let loose in Manhattan and that killed 140,000 people within a week.[40] In May 2000, the administration conducted its biggest disaster exercise ever, Operation Topoff, with 20,000 high-ranking government officials from 35 agencies enacting a combination of biological, chemical, and computer attacks on three American cities. In one of the simulated attacks, the focus was a single perpetrator introducing plague *(Yersinia pestis)* into the ventilation system of a public building.[41] For a number of reasons, the exercise ended in predictable chaos. It was quite clear that in a city like Denver, health services and disaster control would quickly collapse in the face of a major infectious disease breakout, whether natural or malicious. Moreover, the authorities would be incapable of controlling the situation, which says something both about the nature of such events and about the dismal situation of American public health.

Some analysts drew entirely different troubling conclusions

from the Topoff exercise: the chaos was simply the result of frenzied, overfinanced, yet totally uncoordinated antiterrorism efforts on the part of various agencies and authorities that since the end of the Cold War had requested (and received) huge amounts of money for the fight against "terrorism" as the *new* threat. Already at the end of 1997, the General Accounting Office criticized the government's antiterrorist activities for being "based not on a rigorous assessment of threat but rather on the potentially infinite vulnerability of American civilians at home and U.S. armed forces abroad"; there was no way to effectively oversee the allocated monies.[42] A growing number of people realized that there had actually been very few attacks using biological agents, the threat of terrorist incidents was declining, and the few events that had occurred had not involved genuinely dangerous substances.

Indeed, there was little point in talking about biological weapons of mass destruction, because their production, handling, transport, and use would be far too costly, technically too difficult, and extremely dangerous even for well-organized and well-financed independent groups. That much was demonstrated by the Aum Shinrikyo sect's 1995 attack on Tokyo.[43] The likelihood of the United States being hit with such weapons was therefore practically zero, as FBI director Louis Freeh testified in 1999 before Congress.[44] The assertion that it isn't a matter of *whether* a bioterror attack will occur, but *when*—as Clinton never tired of repeating—simply had no statistical justification. And so? During that period, journalist Richard Dreyfus noted, tongue in cheek, that disaster exercises such as Topoff were enjoying increasing popularity: 32 in 1996, 53 in 1997, and 116 in 1998![45] He did not give the figure for 1999; in 2000, as mentioned, Topoff was the most extensive, with fairly predictable results.

Then Clinton left the White House, and in January 2001 the Bush presidency took over. Bush's first speech to Congress on February 27, 2001, was succinct and not yet up to the mark of official threat discourse: the threats to U.S. security "range from terrorists who threaten with bombs to tyrants in rogue nations intent upon developing weapons of mass destruction."[46] But Bush was a quick study. On May 1, 2001, speaking before students at the National Defense University in Washington, he explicitly addressed the national security situation for the first time. Several states not only intended to develop nuclear, chemical, and biological weapons but now actually possessed them. All were working on ballistic missile projects, which constituted a threat to the United States—"states for whom terror and blackmail are a way of life."[47]

What followed was a clear rejection of the old concept of nuclear deterrence. Bush outlined a new, two-part strategy. He agreed with Clinton on the issue of constructing an antimissile defense system. And he recapitulated in somewhat vague language the caveat invoked by Clinton and defense secretary William Cohen that the United States would take unilateral measures against rogue states whenever such states threatened the U.S.'s "vital interests" (specified in Cohen's annual report of 1999 as "uninhibited access to key markets, energy supplies, and strategic resources").[48] Bush was careful not to be too explicit, and stressed instead active nonproliferation of weapons of mass destruction. Security against the new dangers threatening the country could no longer be only defensive but would also include offensive and unconventional approaches. The moment had come "for the world to rethink the unthinkable"—normally we rethink the thinkable or think the unthinkable—"and to find new ways to keep the peace," namely, "a broad strategy of active nonproliferation, counterproliferation and defensive forces."[49]

The fact that the Bush specifically mentioned antimissile

defense in transitioning to this passage should not be made too much of. The new administration was already planning a change to United States defense policy that involved in particular both defensive approaches to weapons of mass destruction and active nonproliferation, as the neoconservative masterminds of the Bush administration and William Kristol, editor of the Washington *Weekly Standard*, had been advocating for years. Put more precisely, the idea was to have a policy of active military intervention, for which nonproliferation of weapons of mass destruction promised at least to provide a rationale.

This speech was Bush's last on the question of a possible terror threat to the United States up to September 11, and given its purely hypothetical future scenario, it received little attention. Not that the administration abandoned the topic. Quite the contrary. On July 22–23, 2001, another big disaster exercise took place ten miles from Washington, at the Andrews Air Force Base. Once again, the exercise, called Dark Winter, involved a simulated bioterrorist raid on three American cities in which (once again) the public health service response to the attack broke down. The exercise was designed around a putative Iraqi smallpox virus of the weaponized variety originating from the former Soviet bioweapons program and purveyed by terrorists. Directed by the Johns Hopkins Center for Civilian Biodefense Strategies, the exercise script was based on intelligence reports stating that besides Russia and the United States, North Korea and Iraq also possessed smallpox, as reported in the *New York Times* of June 13, 1999 (though the article also cited the thin proof for and common doubts about the allegations).[50]

The antiterror exercise at Andrews Air Force Base was based on the following premise: The previous month, Russian authorities had arrested "Yusuuf Abdul Aziiz," a close friend of Osama bin Laden and senior Al Qaeda operative for attempting to procure fifty kilograms of plutonium and various weapons-grade pathogenic agents on the black market. For purposes of the exercise, the U.N. weapons embargo had ended, Iraq had returned to full-scale production of bioweapons and moreover had deployed its troops along the Kuwaiti border. The exercise included various options for an American move against Iraq, provided only in very sketchy form on the Internet.

At the same time, participants are informed that the Middle East supplies 26 percent of the 8.46 million barrels of oil the United States imports every day and that the Strategic Petroleum Reserve will suffice for only 54 days. A defector claims that the Iraqi secret service is behind the attack—and finally, at the end of the "briefing," the letters emerge: "The NY Times, Washington Post, and USA Today receive[d] anonymous letters demanding the immediate removal (one week) of all US forces from Saudi Arabia and all war ships from the Persian Gulf. Failure to comply will result in renewed attacks on US, which will include anthrax, plague and small pox. Each letter also contained a genetic fingerprint of the smallpox strain matching the fingerprint of the strain causing the current epidemic."[51] Not the spores or the virus themselves, just their "fingerprints"—but still.

What happened three months later is history, and it is no wonder that today the United States seriously considers a future bioterror attack to be inevitable.[52] Following the anthrax letters of autumn 2001, the National Institutes of Health budget allocation for antibacterial and antiviral agents alone increased from tens of millions of dollars to about $1.5 billion.[53] In February 2003, President Bush titled the ensemble of research efforts in defense of the anticipated "bioterror" Project BioShield and gave it highest priority.[54] Citizens could prepare themselves for the event with a growing number of guidebooks that began to appear on the bookshelves at the same time as the attacks in early October 2001—as though the letters were an afterthought.

A few examples: Angelo Salvucci, *Biological Terrorism: Responding to the Threat* (October 9, 2001); *21st Century Anthrax Digest: Government Information on Biological Warfare and Bioterrorism: Symptoms, Vaccines, and Treatment* (U.S. Government, October 15, 2001); *Anthrax: A Practical Guide for Citizens—What You Should Know, What You Can Do, and How It Came to This* (by the Parents' Committee for Public Awareness, October 17, 2001); *Bioterrorism after the Anthrax Attacks: Complete Revised Guide to Biological Weapons and Germ Warfare—Anthrax, Smallpox, Medicines, Treatment, Preparedness, White House, Homeland Security, CDC, HHS, FDA, NIH, Military Manuals* (U.S. Government, CD-ROM, 2002); Terri Rebmann et al., *Clinical Description and Epidemiology of Bioterrorism Agents: Anthrax, Smallpox, and Plague* (CD-ROM, April 1, 2002); Bill Frist, *When Every Moment Counts: What You Need to Know about Bioterrorism from the Senate's Only Doctor* (March 2002); Kimberly Lindsey, *Survive Anthrax: How to Prepare Your Family for an Anthrax Terrorist Attack* (August 2002); Dotty Heady, *"The Check Is in the Mail" . . . Along with An-*

thrax, Small Pox, and Your Latest Contest Offer Protecting against Contaminated Mail (September 16, 2002); Garry Null and James Feast, *Germs, Biological Warfare, Vaccinations: What You Need to Know* (February 2003); not to mention for builders, Wadyslaw Kowalsky, *Immune Building Systems Technology* (September 26, 2002), "your complete guide to building ventilation and air treatment systems design! Immune Building Control Systems takes a comprehensive approach to the protection of buildings against biological pathogens" . . . The list goes on.

4

What Is an Author?

More than three decades ago, on the heels of Roland Barthes, Michel Foucault asked what it means to label a person as the author of the work that is attributed to him.[1] Traditionally, the author was ultimately responsible for his statements. It is the author himself that gives a work, a book, or a collection of letters its coherence and unity. Authors "know" what they are doing, and the "meaning" of their text cannot be unlocked without knowing their intention—or, at least, that is the conventional way of looking at it. As I said, Foucault was not the only one to question this assumption at the end of the 1960s. Barthes had already bluntly announced the "death of the author." And Julia Kristeva disavowed the apparent uniqueness of texts by positing connections between authors and their readers in a web of intertextual references, cross-references, other voices, and allusions. Thus, even texts of famous authors lost their coherence, their distinctiveness, and their originality.

It would be easy to read Richard Preston's *Cobra Event* in this way. Or to put it more precisely: it is the only way to understand the novel. Preston exercised what Foucault calls an "author function" by lending his adroit words to various high-ranking functionaries, officers, and advisors of the Department of Defense and the intelligence agencies, speaking with virologists, allowing the secretary of the Navy to dictate his plot, and by reworking classified material to which he had access.[2] Indeed, under closer examination, Preston's story separates into fragments of discourse (more than fragments, actually) that did not just happen to drift from the shadowy, inaccessible world of military secrets into the real world of science fiction novels. At first blush, Foucault's author function might seem to be at odds with the notion of intertextuality, but he had a reason for asserting it. Among other things, the modernist author represents "the principle of a certain unity of writing"—for example, the unity of narration that enabled Richard Danzig to deliver to his president a message the president would not otherwise have heard.[3]

To accept Foucault's author function is to concede that even if the author has ceased to be the unique control center of a text's meaning, the social consequences of imagined attribution guarantee that the author—and he alone!—still must answer for his text. For people like Danzig, what could be better than authors who relieve them of words that they cannot take responsibility for? Can secretaries of the Navy really invent hybrid smallpox viruses that cause russet-haired seventeen-year-old girls to devour themselves as their brains begin to dissolve under Russian attack? Or may they state, without proof, that Iraq is building new attack weapons with genetically modified viruses? They can try, of course, but what is one national security hypothesis among many?

Compared with the relatively simple case of Richard Preston's novel, the question of the author of the anthrax letters is more complicated. We know that the letters contain "handwriting" and that, consequently, even as anonymous messages they most likely carry the unique and identifiable trace of their origins. In contrast to books, we naturally assume that every letter has an individual sender, someone who eschews the hide-and-seek of novelists and who takes responsibility for what he has written. So we find anonymous letters disturbing—or, in the case of a familiar scam, we just throw them in the wastebasket.

But the deadly spores had the effect of causing us to reflect again on the author as we once did and to take his message seriously. It hardly matters whether several people wrote the letters, filled them with spores, addressed the envelopes, and took them to the post office, or whether the letters were the work of a single perpetrator (both hypotheses exist). From the point of view of criminal justice, there is a lot at stake in establishing the identity of this author and his or her co-authors. Ironically, even Foucault admits that the author as historical figure appeared once it was clear that writing books or articles—or letters—was a risky business: "Texts, books, and discourses really began to have authors (other than mythical, "sacralized" and "sacralizing" figures) to the extent that authors became subject to punishment, that is, to the extent discourses could be transgressive."[4] Even if the "poison" of these texts was entirely metaphorical, it was important to know who had spread it.

The question is perfectly legitimate. But it does nothing to resurrect the imaginary unity and simplicity of the author as the "real" source. To put it more simply, finding the perpetrator of the letters is only one part of the task. The other is to understand which discourses and which contexts led to his being the author/perpetrator. Just as we have to reconsider the naive concept of the author as the sole creator of books, here we must frame the question of the perpetrator in the context of cultural studies (leaving the purely criminal aspects to others). Let us then consider the five anthrax letters as a brief "work," in the same sense as Foucault speaks of the work of an author. And let us dare to assume that the anthrax letters do not have an individual author but rather that they force us to consider the possibility of many "perpetrators" or "authors." What discursive and political conditions helped these letters along and even to some extent gave rise to them?

To what ideas, suggestions, or phantasms, to what schemes or motives did this author lend his terse, lapidary expression? And in what discourses did he detect the demand to do "it"? What led him to feel he had the authority to send the letters? Like everyone else, all I can do is speculate. Nevertheless, these questions fall into four areas that we must explore to answer our questions. First, the American biodefense program since the 1990s; second, the question whether the United States was "warned" of a terror attack on the scale of September 11 or could at least have imagined such an event; third, the fevered anticipation of a bioterror attack immediately following 9/11; and fourth, the question of phantasm: Why did the perpetrator send anthrax in his letters rather than arsenic, plastic explosives, or ricin? Why microbes, of all things?

Secret Bioweapons Programs

Weaponized anthrax spores are a weapon of mass destruction, prohibited worldwide under the Biological and Toxin Weapons Convention of 1972. Already by the end of the 1960s, anthrax was no longer officially part of the defensive arsenal of the United States. Shortly before September 11, however, the suspicion arose—and was confirmed by the Pentagon—that under the Clinton administration both the CIA and Defense Department had revived research on anthrax with the aim of developing biological agents for allegedly purely defensive purposes. On September 4, 2001, the *New York Times* reported that according to some U.S. government officials, this program of secret research on biological weapons tested the limits of the global treaty.

The projects—whose scope and development were sometimes kept from the Clinton White House—included reverse-engineering Soviet germ bombs.[5] At issue were so-called bomblets, that is, minibombs transported like cluster bombs by a rocket warhead and scattered over a wide target area, where they then explode individually. Richard Preston described these bomblets in detail before a joint Senate committee meeting on terrorism and intelligence in 1998: "When the warhead reached a certain height over the ground, it burst apart, and bomblets full of smallpox would fly off [in] all directions. The bomblets were egg-shaped and made of aluminum, and were about the size of small melons. They would pop open with a soft wound, and powdered smallpox (or Marburg, or Black Death) would disperse in the air over the city, almost instantly becoming invisible. The powder is very fine."[6] The secret biodefense program Clear Vision, begun in 1999 under the direction of the CIA, built such bombs to study their effectiveness—allegedly according to a single existing Soviet blueprint.

In the meanwhile, attempts by the intelligence community to procure any of these bomblets on the black market came to nothing. Where terrorists and rogue states could avail themselves at will, the American agents were left empty-handed. In other words, these bomblets that can scatter plague and smallpox exist only in Richard Preston's narration and in American weapons laboratories. No matter. Barbara Hatch Rosenberg pointed out the fallacy of such programs: "Whether a bomblet emitted 5 or 50 percent of its deadly cargo was irrelevant if the germs were sufficiently potent" to unleash an epidemic; and in any event, military planners "had to assume the worst." So why (re)produce the bomblets?[7]

In parallel with this project, a bioweapons facility was constructed for producing, among other things, "anthrax simulants"—another form of anthrax that is relevant to our topic.[8] The point in building this little factory was simply to demonstrate that such a plant could be assembled using commercially available components. Yet it is not completely clear that production was limited to the imitation agent. As early as autumn 2001, the test factory at the Dugway Proving Ground in Utah was named as one of the few probable sources for the anthrax contained in the five letters—not the anthrax simulant, but the real thing. The reason, as was soon revealed, was that it was one of the few plants where spores could be milled to a grade finer than five microns, using technology based on five secret patents and that only very few U.S. scientists would have had access to.[9]

The spores in the poison letters were all the same Ames reference strain, named for the town of Ames, Iowa, where this variety of *Bacillus anthracis* was first isolated.[10] The strain was originally bred in the U.S. Army Medical Research Institute of Infectious Diseases (USAMRIID) in Fort Detrick, Maryland, and is now used by many laboratories. But the anthrax in the letters matched a subtype of the Ames strain—the one used in experiments at the Dugway Proving Ground. No less than Richard (Dick) Spertzel, a retired colonel in the U.S. Army and head of the U.N. Biological Weapons Inspection Team in Iraq, expressly rejected the thesis that an isolated madman like the Unabomber could have sent the anthrax letters. The anthrax was of superior quality, which implied exclusive access to special military technology. Appearing before a Senate committee on December 5, 2001, Spertzel explained: "The quality of the product contained in the letter to Senator Daschle was better than that found in the Soviet, US or Iraqi program, certainly in terms of the purity and concentration of spore particles."[11]

The Clear Vision program for the development of "defensive" bioweapons was a classified project under the Clinton administration, that is, a program subject to the fairly substantial concerns of officials in the White House that the bioweapons treaty not be undermined. According to Miller, Engelberg, and Broad, in its final months the Clinton administration basically lost interest in Clear Vision and pulled back financing on the project. Only with the start of the Bush government did work resume, for instance, on very potent anthrax agents geared to offensive warfare.[12] After September 11, to be sure, as the Austrian bioweapons expert Georg Schöfbänker remarked wryly, all this "quickly fell victim to collective amnesia by the censors as unpatriotic."[13]

But not for long. As early as December 2001, the first reports of the U.S. bioweapons program emerged, and on July 2, 2003, the *New York Times* divulged a further aspect of the program. According to the report, a bioweapons expert and former employee of USAMRIID in Fort Detrick had since 1999 been working on developing and building a mobile germ plant for the Pentagon. This mobile laboratory contained all the equipment and components needed for breeding germ agents and weaponizing them: "a fermenter, a centrifuge and a mill for grinding clumps of anthrax into the best size for penetrating human lungs."[14] In spring 2003 the FBI suspected that the mobile unit had been used to manufacture the anthrax spores for the letter attacks, but failed to turn up any trace of spores in it.

Both the bioweapons expert and the Pentagon insisted that the purpose of the lab was simply to familiarize the Army's elite Special Operations Delta Force with such units in case they ran into something similar in Iraq. The *Times* pointed out, however, that the U.S. armed forces had already possessed mobile bioweapons production plants fifty years previously. The idea was that, in the event of destruction of fixed plants by a Soviet atom blast, mobile units could still manufacture sufficient quantities of agents to annihilate Soviet cities. Today the scene is completely different, as Colonel Bill Darley, spokesperson for the United States Special Operations Command, underscored in July 2003: "We are not growing anthrax or botulinum toxin. None of this equipment is functional. It looks like—it is—the real stuff, but it's nonfunctional."[15]

At any rate, all these secret weapons programs—officially, biodefense programs—were the reason that in July 2001 the United States rejected the protocol of the so-called Ad Hoc Group of the Bioweapons Convention.[16] That in turn set up the failure of a conference called in Geneva in December to sign the renegotiated protocol for the treaty. On July 25, 2001, U.S. Ambassador Donald A. Mahley offered a muddled explanation for the refusal: "For instance, we believe that increased capability to resist disease, among other things, lowers the probability that a biological weapons attack would be successful, and therefore in some ways lowers the desirability of biological weapons for a potential proliferator or for a terrorist."

A journalist asked: "A follow up, if I may. What do you mean by a higher resistance to disease?" Mahley responded: "One of the things that compound the effectiveness of a biological weapons attack, if one were to occur, is a general state of debilitation or lack of disease resistance on the part of the target population. If one corrects that, then one automatically lowers the potential effectiveness and the immediacy of spread of biological weapons attack and we think that is also, while not a biological weapons convention matter itself, nonetheless something which helps in terms of the biological weapons threat situation around the world."[17] Did the Bush government actually think that improved nutrition would protect the American people from bioterror? Unlikely. In any event, the U.S. government was clearly opposed to the "challenge inspections" foreseen in the protocol—on-site inspections in any of the 143 signatory states, including the United States, on suspicion of violation—out of fear of industrial or state espionage.

Since the days of the Clinton administration, the United States had been arguing as a matter of principle that genuine verification was just not feasible. Consequently, in Mahley's words, the government relied more on "codes of ethics and other kinds of activities that would be enduring means of trying to remind people of the fact that the biological weapons are not things to do."[18] But after September 11 and particularly after the anthrax letters, Bush decided to proceed aggressively with the "defense" research on bioweapons begun under Clinton—"to greatly expand the number of such clandestine projects," as the *Times* reported—and he wanted no foreign monitoring or control of these activities.[19]

These references to the efforts, especially of the Bush government, not to make accountable or public its own biodefense research—which naturally also generates knowledge for offensive applications—takes our investigation into an very murky area that I am at pains to elucidate. The difficulty is manifest in the latest (published) reports of scientific research on the anthrax powder used in October and November 2001. On November 28, 2003, journalist Gary Matsumoto published a long article in *Science* magazine on the status of the FBI inquiry into the anthrax letters. The article confirms Richard Spertzel's assessment that the quality of the spores in the letters to the two senators bespeaks technological savvy far more advanced than any known bioweapons program. What especially amazes experts is not only the degree of refinement of the spores but the fact that they were electrostatically charged and treated with silicon nanoparticles as well as polymerized glass—a material not previously encountered in this connection. These additives prevented the spores from clumping and allowed them to stay airborne in test tubes for long periods of time.

To date, so far as we know, only the Battelle Memorial Institute in Columbus, Ohio, and its industrial spin-off, BattellePharma, possess the technological expertise to develop a self-dispersing powder of this quality. Batelle scientists produced Ames-type anthrax aerosol for the U.S. Army, so Battelle was included in the FBI investigation—thus far to no avail. According to Matsumoto, the technology used for these extremely fine particles is so cutting edge that bioweapons experts have concluded that nobody can procure anthrax of this quality without substantial state support. Thus one cannot strictly rule out the regime of Saddam Hussein as a sponsor for the perpetrator, although the *Science* article states unambiguously: "If the Senate anthrax powder did in fact have these refinements, its manufacture required a unique combination of factors: a strain that originated in the United States, arcane knowledge, and specialized facilities for production and containment. And this raises the discomforting possibility that the powder was made in America, perhaps with the resources of the U.S. government."[20]

But exactly who and where? The anthrax laboratories at Fort Detrick had allegedly been decrepitating for decades, and results of tests at the Dugway Proving Ground showed that the prefab laboratory could not produce anthrax of comparable quality. Spores could be ground to five microns, but they clumped together and sank immediately. So where did the anthrax in the Daschle and Leahy letters, in particular, come from? The question remains a mystery. In February 2004, *Time* magazine stated: "Still, it's worrisome to know that anyone is sending lethal substances through the U.S. mail—and getting away with it. The FBI has spent 251,000 man-hours on the anthrax case, conducted 15 searches, interviewed 5,000 people and served 4,000 subpoenas—without an arrest."[21]

Presentiment and Foreknowledge

In April 2002, journalist David Tell of the neoconservative, pro-government *Weekly Standard* sought to justify the "suspicion" that terror groups such as Al Qaeda were most likely equipped by governments such as Saddam Hussein's to carry out anthrax attacks against the United States.[22] And on September 8, Vice President Cheney repeated this allegation in the form of a revealing speculation.[23] Although up to the war in Iraq the media successfully linked anthrax letters, "weapons of mass destruction," and Sadam Hussein, on closer scrutiny these connections do not hold up. The FBI and most observers are now proceeding on the assumption—not least on the basis of elaborate handwriting and textual analyses of the letters and envelopes—that the perpetrator is an American from the bioweapons research community.[24]

Assistant FBI Director Van Harp, head of the anthrax probe (which he self-consciously code-named Amerithrax), is credited with nailing down at least the geographic aspects of the investigation. The list of potential perpetrators that resulted comes to no more than twenty names. The November 2003 *Science* article punctured the theory that the perpetrator or perpetrators acted alone or at least without access to state support—that is not to say they acted on behalf of a government. Moreover, it is probable that the perpetrator(s) has close connections to right-wing extremists in the United States. Since well into the Clinton presidency, these groups had campaigned aggressively (at least rhetorically) for the use of violence against Democratic Party members. Daschle had already been a preferred goal of such attacks for a long time. The conservative weekly newspaper *Human Event* branded him "Osama's Enabler in Congress." In fact, the anthrax letter containing the line "You cannot stop us!" was sent to his office precisely when Congress—following an initial truce in deference to 9/11—had resumed its habitual political bickering. Moreover, Daschle and Leahy, in defiance of the explicit opposition of the White House, demanded an independent commission of inquiry into the airplane attacks. The story doesn't stop there.[25] But in any event, what was clear at that point was that there had basically been two *separate* anthrax letter attacks: one directed to media people to add the prefix "bio" to the terror of September 11, and a second, using very fine, very potent spores in a more than thousandfold lethal dose, targeted directly at the two senators.

Assuming that the perpetrator is an American who worked (or is working) in the bioweapons program, and recalling that the first letters were mailed off only seven days after September 11, two troubling questions arise. Bioweapons expert Barbara Hatch Rosenberg formulated the generally accepted theory that the anthrax attacks were not an impulsive act, triggered by the TV images from New York, but an attack carefully planned beforehand. Nobody obtains high-quality anthrax just like that. Hatch Rosenberg is confident that the anthrax "was already in the hand[s]" of the perpetrators, and "the attack largely planned . . . before Sept 11."[26] But who would plan such a thing ahead of time—and why? According to Hatch Rosenberg, the perpetrator was only awaiting the opportunity for a terrorist attack to "throw suspicion on Muslim terrorists" and deflect attention from himself.[27] The goal might well have been to alert the government to the need of investing more money into the fight against bioterrorism. If that hypothesis is correct, the gamble was certainly successful.

Nevertheless, the question arises how likely it is that someone working in the U.S. bioweapons program kept a supply of anthrax powder at home for (possibly) many years awaiting an eventual "Muslim" attack on the United States. With the exception of the first bomb attempts on the garage of the World Trade Center in 1993, such attacks were unheard of in America before September 11. Our author must have waited long and patiently. Or did he sense, or even know, that a major "Muslim" attack was in the works? What do such presentiments have to do with feeling oneself called to do "it"?

Sensed? Known? The official line remains that 9/11 caught

America totally unawares, and that afterward, the intelligence agencies—especially the CIA—were taken to task for their alleged inability to foresee and prevent the events. Let's stick with the perspective of our presumed author and the question of what might have motivated him to write the letters, and let us focus on two aspects of this complex problem. We might, for example, ask the question directly: Was there any forewarning that our bacteria-breeding, spore-grinding author could have gotten wind of through his (possible) connections with America's secret bioweapons research program? Alternatively, we can pose the question in a much more general way: Could the government, the intelligence agencies, and the Defense Department have anticipated such a strike, or was that beyond the imaginable? And above all: In terms of September 11, what was the role of the preparatory exercises for possible bioterrorist attacks?

For the moment, let us consider the question about the warnings. So as not to completely lose the thread of my thinking, I will limit myself to the most relevant indicators. To put it simply, not only was there a whole series of very concrete warnings as early as summer 2001, but the U.S. intelligence agencies had to some extent been pursuing the activities of the September 11 terrorists for years. The most significant warnings are well documented on the CNN website (the synthesis, as it were, of the American mainstream media). Of particular note is an eleven-page intelligence memorandum dated August 6, 2001, and submitted to President Bush, regarding a planned Al Qaeda attack on targets within the United States, in which hijacking of airplanes would play a central role.[28] Corresponding substantive warnings and information about such plans were submitted to the intelligence agencies beginning in 1995 and particularly frequently in summer 2001.[29] But according to official sources, these agencies—reputed for sloppiness—paid no attention to them, although Al Qaeda had long been considered a major hostile threat. The question is, which warnings, memos, and reports regarding which other organizations and enemies in the years and months leading up to September 11, 2001, were ranked so much more important that unfortunately all tips related to Al Qaeda activities ended up buried in the to-do box?

Since the middle of the 1990s FBI agents in Arizona had been systematically observing Arab flight school students and had identified Zacarias Moussaoui, the so-called twentieth hijacker of September 11.[30] It is also a fact that FBI headquarters inexplicably failed to act on warnings transmitted by its own agents, although previous warnings had been taken seriously, for instance, in connection with the Olympic Games in Atlanta.[31] More amazing still: The *Washington Post* ascertained that two of the culprits, Khalid al-Mihdhar and Nawaf al-Hazmi, who lived unchecked in San Diego and were even listed in the telephone book, appeared on the FBI's terrorist watch-list and were monitored by its agents. Yet no one prevented them from buying a first-class ticket, boarding a plane, and hijacking it.[32] And if that weren't curious enough, in this same connection, there was a report in *Newsweek* and *msnbc news* (but not followed up by the other media) that in the 1990s five of the hijackers had received training in secret "U.S. military installations."[33]

Finally, and most remarkable of all, there is the case of Mohammed Atta. He was followed for years by secret services in the United States, Egypt, and Germany, and his link to Al Qaeda was very well known—without his freedom of movement being constrained in any way, as Patrick Martin states, based on American mainstream media.[34] At the end of August 2005, a second secret service officer acknowledged that, in the course of the intelligence program Able Danger, Mohammed Atta had been identified in early 2000 as leader of a planned operation: "'My story is consistent,' said Captain Phillpott, who managed the program for the Pentagon's Special Operations command. 'Atta was identified by Able Danger by January–February of 2000.'"[35]

Briefly, and without further belaboring the point: The claim that the American government was totally surprised by the attacks does not seem to hold water. More probable is that the U.S. government had at least an inkling that "something like that" would happen before long. Regrettably, and for reasons that we can only speculate about, the government failed to take proper defensive measures. In line with this hypothesis was the well-known, well-documented (on the Internet), and altogether astonishing fact that for well over an hour, the four airplanes were venturing way off course—180 degrees off, to be precise.[36] All that time, they were closely tracked by both civilian and military flight-control radar, without any action being taken against them (although something was obviously malignantly amiss), and without interceptor aircraft being sent up, for example, from Andrews Air Force Base, not ten miles from the Pentagon (although this is U.S. Air Force mandatory procedure and is routine in the case of even much smaller flight path deviations—not to mention four simultaneously hijacked planes!), and without shooting at the airplanes, though it was clear that they were carrying out terrorist attacks, at the latest after United Airlines flight 175 whammed into the second World Trade Center tower. Some people asked the obvious question: "Were they on vacation or what?"

As stated in the official 911 Commission Report chaired by Thomas Kean, the North American Aerospace Defense Command (NORAD), under the command of General Ralph Eberhart, is responsible for defending U.S. airspace. NORAD's tasks are clearly spelled out: "NORAD's mission is set forth in a series of renewable agreements between the United States and Canada. According to the agreement in effect on 9/11, the 'primary missions' of NORAD were 'aerospace warning' and 'aerospace control' for North America. Aerospace warning was defined as 'the monitoring of man-made objects in space and the detection, validation, and warning of attack against North America whether by aircraft, missiles, or space vehicles.' Aerospace control was defined as 'providing surveillance and control of the airspace of Canada and the United States.'"[37] According to the statements made in the Kean report, the Federal Aviation Administration (FAA) was supposed to inform the Pentagon. This the FAA did, but simply too late, and consequently NORAD failed to intercept the hijacked airplanes.

David Ray Griffen, professor emeritus at the Claremont

School of Theology, has analyzed the 9/11 Commission Report page for page and rejects as disingenuous the explanation of the NORAD breakdown.[38] The FAA clearly informed the Pentagon and NORAD on September 11, 2001, before the crash of the first plane into the WTC. But somehow NORAD did not react, although from that point up to the crash of the third jet into the Pentagon, another full hour had gone by. Griffin refers to a statement by General Eberhart, "reported in October 2002: From the time the FAA senses that something is wrong, 'it takes about one minute' for it to contact NORAD, after which NORAD can scramble fighter jets 'within a matter of minutes to anywhere in the United States.' Part of the reason they can get anywhere within a matter of minutes is that, according to the US Air Force website, an F-15 routinely 'goes from scramble order to 29,000 feet in only 2.5 minutes,' after which it can fly 1,850 miles per hour."[39]

The novelist Tom Clancy, too, was aware of this fact. In *Rainbow Six* (1998), he describes the dramatic moment when bioterrorists with four airplanes(!) took off from their base in Kansas—these four pilots did, however, officially announce their flights—in order to leave the United States for an undisclosed target. John Clark, CIA agent and commander of the antiterror unit Rainbow Six, wanted them tracked. His boss, CIA director Ed Foley, brought in the Air Force to chase after these de jure compliant flights with four fast, long-range Gulf Stream jets, an AWACS reconnaissance plane, and an E-3B Sentry tanker plane. While Foley was still on the phone, his Air Force interlocutor at the other end called "the North Atlantic Aerospace Defense Command in Cheyenne Mountain, which had radar coverage over the entire country . . . ordering them to identify the four Gs. That took less than a minute, and a computer command was sent to Federal Aviation Administration to check the flight plans that had to be filed for international flights." An already airborne AWACS was redispatched to its new task, and the tanker plane needed for the long tracking "got the word fifteen minutes later." The tracking posed no problem: "The steer from Cheyenne Mountain made the tracking exercise about as difficult as the drive to the local 7-Eleven."[40]

Because the Kean report did not really go into NORAD's unfathomable behavior regarding September 11, and in answer to the central question regarding the incredible failure of air defense referred to the alleged failure of the FAA, U.S. journalists took a closer look at NORAD's response. Michael Ruppert discovered that on September 11, NORAD carried out a whole series of military training exercises known as "war games," including—astonishingly—a simulated terrorist attack on the United States using airplanes. Ruppert reported that the Kean Commission had also summoned General Eberhart to its last public hearing on June 17, 2004. He was asked about the war games and the reasons for them. Eberhart confirmed that such exercises took place but declined to say who was in charge of coordinating them. When the commission asked Eberhart whether the war games helped or hurt NORAD's defense, Eberhart said: "Sir, my belief is that it helped because of the manning, because of the focus, because the crews—they have to be airborne in 15 minutes, and that morning, because of the exercise, they were airborne in 6 or 8 minutes. And so I believe that focus helped."[41]

In its final report the Kean Commission includes incidental mention (in a footnote) of the existence of at least one of these war games: "On 9/11, NORAD was scheduled to conduct a military exercise, Vigilant Guardian, which postulated a bomber attack from the former Soviet Union." The question that raises is obvious: "We investigated whether military preparations for the large-scale exercise compromised the military's response to the real-world terrorist attack on 9/11." And the answer? "According to General Eberhart, 'it took about 30 seconds' to make the adjustment to the real-world situation. Ralph Eberhart testimony, June 17, 2004. We found that the response was, if anything, expedited by the increased number of staff at the sectors and at NORAD because of the scheduled exercises."[42] Thirty *seconds* to notice that the four hijacked aircraft were no exercise—thirty seconds, because owing to the exercises, so many of the military surveillance personnel in this sector were occupied with air defense. That, unfortunately, is the stunning contradiction of the official 9/11 version that the commission didn't pursue.

If "the circumstances" for air traffic control for NORAD were particularly good thanks to the war games, and 30 seconds was all it took to distinguish play and reality, it is even harder to understand how everything could go awry on that morning. The terrorists could not be stopped because people supposedly were not ready. So what did NORAD do at the end of thirty seconds? Did the war games cause so much confusion on that day that no fighter jets took off? Or was the point of the games to deliberately flummox NORAD personnel? The Kean Commission did not ask.[43]

These ambiguities are all the more disturbing in that, even independent of the war games, the airplane attacks can hardly be called "surprising." On the contrary: To maintain that the

U.S. government was not prepared for a terrorist attack is not only implausible but a bad joke. Because at the time, apart from the above-mentioned warnings, the U.S. government was spending $800 million per year on early detection and defense against terrorist attacks. Among other things, with this money and since 1998, the government organized more than two hundred antiterror exercises—that is, on average, one exercise per week, year after year. As early as 1993, the Pentagon carried out an exercise in which commercial jets were used as bombs targeted to important buildings in the United States. And on November 3, 2000, antiterror experts even simulated the crash of hijacked airliners into the Pentagon using tabletop models of buildings. Not to mention Tom Clancy's best-selling thriller *Debt of Honor* (1994), in which terrorists plan to crash a Boeing 747 on Capitol Hill. The book's first printing ran to 2 million copies.[44]

Given all this confusion, let us return to our author. Just suppose that he actually does have a military or intelligence service background. Then it would not be so far-fetched to think that he might have had an idea that something was going on. Provided one had the necessary information—or even just a receptive ear for the many rumors then floating around Washington—he might have reason enough in summer 2001 to believe that it soon might come in handy to have some anthrax powder on hand to maximize the "bio" dimension of a terrorist strike. The time was ripe. Except, why should our author do it? What made him think he had to pick up a pen and compose his little missives, sprinkle them with powder, and take them to the post office (taping the envelopes shut, just to be sure)—in the hope that their recipients would soon and with appropriate fanfare notice the difference between the humdrum anthrax hoax letters and the deadly earnest intent of *these?* Why not go fishing, or take the kids on a trip, or write a scientific article?

The question is loaded with mystery. Why should someone spread fear and dread with these spores? Is it not enough—from any conceivable perspective of an attacker or terrorist or Islamic fundamentalist or whoever—to have the World Trade Center towers collapse, some 2,600 people buried under them, and a part of the Pentagon up in flames? What a strike for terror! How could you top that? But for *those under attack,* a critical element was still missing. They were expecting a bioterror attack: poison, disease, gas, as Bush said later in Oklahoma; Marburg or black death, in the words of Richard Preston; the superbug of the Dark Winter exercise, smallpox, or at least anthrax—a lot of anthrax.

The Anthrax Panic *(Avant la lettre)*

The feverish anticipation of a bioterrorist attack started slowly in the years leading up to September 11 and increased sharply from the moment of the World Trade Center attack until it reached the anthrax frenzy. Up to two-thirds of the already cited 200-odd government-level antiterror exercises had figured on biological or chemical attacks. And this despite several government-sponsored studies, as well as a highly critical statement by the General Accounting Office, which concluded unambiguously that, as the *Washington Post* wrote, "bombings, hijackings, and other low-tech missions were far more likely." Experts who suggested the need to prepare for these threats were paid little heed.[45]

It cannot be stated more clearly: Bioterror is a phantasm; nevertheless, 9/11 appeared to confirm our worst fears. In the words of L. Paul Bremer, then chairman of the National Commission on Terrorism and later "civilian administrator" of Iraq, the attacks were felt as a threshold to be crossed on the way to bioterror.[46] On September 30, the *New York Times* reported that already "minutes after two jets slammed into the World Trade Center, the National Guard was mobilized. The Guard has created 29 teams around the nation to aid the response to chemical, biological, and radiological attacks; on Sept. 11, a 22-member unit was ordered into Manhattan to test the air for deadly germs or chemical toxins."[47] At 1 p.m. the CDC in Atlanta sent out a "health alert" for possible bioterror to all the hospitals nationwide.[48] Elsewhere, things had already gone a step further: From the "early hours" of September 11—it would be nice to know when, exactly—White House personnel were given the powerful antibiotic Cipro "before the situation could be fully assessed," Gordon Johndroe, a White House spokesman, acknowledged in June 2002.[49]

Ciprofloxacin is manufactured by the German pharmaceutical giant Bayer and in summer 2001 was the first antibiotic allowed onto the American market by the Federal Food and Drug Administration (FDA) for use in biological attack.[50] The drug is not only a good antidote against anthrax, provided it is detected early enough, but also against simple conspiracy theories. Let us assume that the reconstruction of the events, as sketched below, is more or less accurate (at least for purposes of reflection): The administration had taken note of the warnings of the forthcoming attack and had ensured that it would not be prevented. Maybe it figured that such an attack would involve the nation in a long war on terror that could enable many geopolitical options and could even have interesting political ramifications at home.

But that does not mean by a long shot that the government actually knew what mischief Atta and his sidekicks were up to (nor did it foresee—or could it have foreseen—the collapse of the towers). It is not hard to imagine the apprehension of those playing their dangerous game with the devil: What if the attackers had taken not just box cutters onto the plane with them, but also a Pandora's box filled with anthrax agents or the like? Or what if the co-conspirators located elsewhere in the United States had released such microbes? Anything seemed possible. Hence the National Guard's mobile detector unit, the early CDC warning, and the Cipro in the White House. These were the first indicators of the nascent panic—long *avant la lettre*.

On September 12 at 2 in the morning, a group of five experts—Marcelle Layton, New York's public health commissioner for communicable diseases, Joel Ackelsberg, commissioner of the State of New York for bioterror defense, and three officials of the CDC—met in Layton's office to prepare twenty-nine of the city's hospital emergency wards for the consequences of a bioterror attack.[51] Simultaneously, for the first time in American history, the U.S. National Pharmaceutical Stockpile was tapped, and antibiotics and other medical materials were made available in case of a bioweapons attack.[52] On September 13, under the headline "Vulnerable home front," *USA Today* listed the dangers that now threatened the United States, with "biological weapons" at the top of the list: "Just 250 pounds of the infectious disease anthrax spread [sic!] over the Washington, D.C., area could kill up to 3 million." The United States was "woefully unprepared" for such an attack.[53]

That same day, the Army ordered from the firm Bruker Daltonics a further lot of bioagent detection machines, for which the Army had shortly before signed a $10 million contract.[54] Three days later, on September 16, high-ranking experts at the Center for Civilian Biodefense Strategies of Johns Hopkins University Medical School in Baltimore—where the script for the Dark Winter exercise originated—briefed health secretary Tommy G. Thompson about the widely expected second strike with biological warfare agents.[55] And on September 22, the *New York Times* quoted an editor from the fashion world as saying, "If I were a designer, I would be working with NASA and perfecting clothes impervious to anthrax. I'd try to design clothes to solve the problems of our troubled world."[56]

There is no shortage of such examples. But the key message was broadcast the same day by the online version of *Time* magazine. As mentioned above, the FBI suspected that "Bin Laden conspirators" had planned to use cropdusting planes to disperse biological or chemical agents. Indeed, "manuals" on cropdusters had been found among the belongings of Zacarias Moussaoui (who had himself been taken into custody). (The "manuals" later turned out to be a file on Moussaoui's laptop—he had once searched "cropduster" in Google.) So beginning September 16, cropdusters all across the country were immediately grounded.[57] At the same time, the media peddled the rumor that a cropduster mechanic in Florida had told CNN that a group of Arabs, including one who looked like Mohammed Atta, had inquired about the carrying capacity of cropdusters for agricultural use.[58]

These semiofficial rumors once again roused people—especially New Yorkers—to a state of high anxiety. "Fears of anthrax hung in the air over parts of the Lower East Side," reported the *New York Times* on September 23.[59] And four days later columnist Maureen Dowd wrote, "Women I know in New York and Washington debate whether to order Israeli vs. Marine Corps gas masks, and half-hour lightweight gas masks vs. $400 eight-hour gas masks, baby gas masks, and pet gas masks, with the same meticulous attention they gave to ordering no-foam-no-fat-no-whip lattes in more innocent days. They share information on which pharmacies still have Cipro, Zithromax, and doxycycline, all antibiotics that can be used for anthrax, the way they once traded tips on designer shoe bargains. They talk more now about real botulism than its trendy cosmetic derivative Botox."

The article quoted a doctor in Manhattan as saying that her patients were "taking their little black Prada techno-nylon bags and slipping in gas masks for the couple, Cipro, a flashlight, a silicone gel tube—you smear the silicone on your skin so hopefully it doesn't absorb the spores as fast. It's truly scary."[60] Even *Times* reporter Judith Miller—forced to leave the paper in October 2005 over her blatantly biased reporting of the leadup to the Iraq war—weighed in: "It's the ultimate freakout." It comes as no surprise, then, that she muted her criticism, acknowledging reason to fear "Muslim martyrs willing to be infected with smallpox or Marburg, a cousin of ebola, who could then walk around our malls and cause an epidemic."[61]

Many so-called bioterror experts now warned of looming threats of smallpox or anthrax, and the media reported sellouts of gas masks in New York, runs on handguns and bottled water, and hoarding of Cipro. A pharmacist told a reporter for the *New York Times:* "We can't keep it in stock. It started the day after the World Trade Center, with a few prescriptions, and now there's more and more. I usually keep 100 tablets in stock, but this time I ordered 3,000 and sold out. One person bought 1,000 tablets."[62] The anthrax frenzy wasn't limited to an ignorant public. In the last week of September, on *Larry King Live* on CNN, Senator John Kerry urged parents to have their children immunized against smallpox and anthrax. The American Academy of Pediatrics posted information about the two diseases on its website because its members were being inundated with calls from worried parents.[63] All over the country, access to dams and reservoirs was being blocked, in particular New York's reservoirs, which were monitored by helicopter. This despite explanations by "experts" that water is not a good way to spread anthrax—even "a Boeing 767 laden with anthrax" crashing over a reservoir would do little harm.[64] Health Secretary Thompson exhorted the pharmaceutical and biotechnology industries to produce more vaccines against biological agents.[65] The Pentagon financed the deployment of the Michigan National Guard to provide security for BioPort Corporation, in Lansing, Michigan, the only firm nationwide capable of manufacturing an anthrax antibiotic—though no terrorist threat had been made against BioPort.[66]

As the news of the cropduster flickered over TV screens, the first anthrax letters were already on their way or had just reached the mail distribution centers of the media companies they were addressed to. I hesitate to say whether that was coincidence or perfect timing. What can be said is that objectively, the cropduster announcement was so absurd as to border on disinformation. Anyone with so much liquid anthrax that they can conceive of using a cropduster to spread it is planning a terror attack whose dimensions dwarf the operational details of 9/11. Gallons of liquid anthrax presupposes large-scale industrial production capacity, a ready supply of money, and very carefully thought out high-tech transport logistics. Nobody has ever claimed that terrorists of whatever stripe have such resources to draw on. A person or group capable of planning bioterror of this magnitude does not have to search for cropduster manuals over the Internet or sound out mechanics on the subject of spraytank capacity. Conversely, someone making inquiries about cropdusters evidences only that he is about as dangerous to the safety of the free world as kids (young or old) keen about fighter aircraft. In any event, it is scarcely coincidental that the cropduster alert played no further role in the legal proceedings against Moussaoui.

Nevertheless, if the cropduster did stir up panic in September 2001, it is probably because it is a fixture of American cultural imagination dating from Alfred Hitchcock's 1959 film *North by Northwest.* In a key scene of what is perhaps the director's best-known thriller, the hero (played by Cary Grant) is attacked by a low-flying cropduster in an abandoned cornfield: a civil aircraft turned into a weapon. It does not matter whether the cropduster news of September 22, 2001, constituted deliberate disinformation or was merely a minor detail blown all out of proportion. The cropduster signifier functioned as the (since Hitchcock) obvious missing link between the low-flying airplanes that in New York and Washington "were converted into weapons" and the impending bioterror. At a point where the fear of anthrax looked like it might peter out, the cropduster spawned an anthrax frenzy as good as the reaction the author hoped to get from the letters circulated only a few days before.

Who Did It?

Indeed, what of our author? We do not want to lose sight of him. How are we to understand his letter writing in the context of these expectations that were an integral part of his letters before he ever took them to the post office and while he was still waiting for their contents to become known? There are mainly two ways of making sense of the connections. On the one hand, authors are independent, sit at home in their garret, and write. They note down everything buzzing around in their head in the way of discourses, ideas, and knowledge from unofficial reports, and also in the way of thoughts shared with friends, newspaper reading, and remembered conversations. Authors are sensitive, in tune with the spirit of the times, and have informants in milieus that are closed to others. They also know how exciting it is, but also sometimes how essential, to publish anonymously when it is not expedient to openly advertise their author function. It may be enough that their friends know who is behind attention-getting letters. Maybe our author actually did his government a favor in surprising it with a gift—but, like all good gifts between friends and acquaintances, one that satisfied a secret wish and, in this case, at the same time a bluntly express desire. It was a real *gift that keeps on giving,* as Clinton said, though he did not mean it that way: a gift with long-term political ramifications tied up with a string of very special signifiers.

The other way of thinking about the connections between

the anthrax letters and September 11 is more sinister. In this version, our author is a two-bit work-for-hire writer, the sort of person who can be tasked with using his art to fabricate a link between the airplane attacks and Saddam's "weapons of mass destruction" because the attackers themselves were too stupid to have thought of it. Maybe they had seen all the Hollywood disaster films but neglected to read *The Cobra Event*. It is even conceivable that our author might have accepted such a job shortly after September 11.

But we still have not answered the simple question: Who did it? Since summer 2002, various experts have suggested that the perpetrator is a "US defense insider," probably "someone with high-ranking military and intelligence connections," according to a linguist with the FBI.[67] In September 2003, *Vanity Fair* published an article by a handwriting expert who believed that a virologist and bioweapons specialist who once worked at Fort Detrick on a project to develop a mobile biolab was behind the letters.[68] However, investigations of the laboratory as reported in *Science* magazine in November 2003 indicated that no one working alone out of a mobile unit would have had access to the necessary cutting-edge technology, especially for making the powder sent to Leahy and

Daschle. The bioweapons expert has taken legal action to clear his name.

The question of the perpetrator, a serial killer with no fewer than five murders on his conscience, remains unanswered.[69] The FBI's apparent ability to mount the most extensive manhunt in its history without getting anywhere found its parallel in the conspicuous thrashing around among the members of Congress charged by the official federal commission with clearing up the events of 9/11. On July 8, 2003, the commission deplored in no uncertain terms the fact that the Pentagon and the Justice Department, apparently at the instruction of the Bush administration, were hampering its efforts, withholding documents, "blocking requests for vital" classified information, insisting that officials be interviewed only in the presence of "minders," and literally, "intimidat[ing]" the commission—in short, the usual Saddam dodge when faced with an investigation.[70] According to an editorial in the *New York Times,* the indignant administration was "acting more like the Soviet Kremlin than the American government."[71]

And the anthrax letters? I dealt briefly above with some of the oddities of the Kean report regarding the incidents of 9/11. Moreover, a careful reading of the report shows that surprisingly, the anthrax attacks of fall 2001 were not only not considered, they were not even mentioned! These attacks are, as it were, not recorded in the official history as represented by the Kean report. This is all the more striking as the report doesn't focus only on September 11, but very broadly delimits the historical period of the investigation. For example, it goes all the way back to the attack on the U.S. Marine compound in Beirut in 1983, and foresees a radical reorganization of the U.S. secret services beginning in 2005. The omission is doubly surprising given that the report repeatedly speculates on various possible connections between terrorists and weapons of mass destruction.

The word "anthrax" shows up only once in the Kean report, and that is in the chapter titled "Al Qaeda Aims at the American Homeland," where it is stated that Al Qaeda had developed a "bioweapons" program in Afghanistan. The report states that "in 2001, Yazid Sufaat would spend several months attempting to cultivate anthrax for al Qaeda in a laboratory he helped set up near the Kandahar airport."[72] Who is this Sufaat? One footnote refers, among other things, to an "interrogation of detainee, Apr. 30, 2003," which, as we know, can also mean torture. The text reads: "Sufaat received a bachelor's degree in biological sciences, with a minor in chemistry from California State University, Sacramento. Sufaat did not start on the al Qaeda biological weapons program until after JI's December 2000 church bombings in Indonesia, in which he was involved."[73]

Microbiologists and bioweapons experts can be forgiven for finding humor in such statements: a simple bachelor's in biology and a smattering of undergraduate chemistry in no way equips a person to develop a "biological weapons program." It is no coincidence that today no credible expert thinks that Al Qaeda has bioweapons or that it could have been behind the fall 2001 anthrax attacks. On the other hand, there is no denying that, in their terrorist zeal, members of Al Qaeda could have explored a broad range of weapons and warfare agents, and in the process also have taken an interest in anthrax, among other things. But, according to experts, the difficulties for laypeople simply to procure a virulent strain of anthrax are virtually insurmountable—to say nothing of turning them into the requisite aerosols. And we should also recall how surprised U.S. troops in Afghanistan were, on entering a building identified by the secret service as a onetime Al Qaeda "bioweapons laboratory," not even to find running water.

Martin Schütz, head of the biology division of the Swiss defense department and former U.N. weapons inspector in Iraq, has a completely different idea of how to make headway in the search for the perpetrator of the letters: "Why doesn't someone ask Richard Preston who the author of the anthrax attacks was? What the hell! It might be Mark Littleberry!"[74]

II

Microbes

5

Foreign Bodies

One question remained unanswered: Why microbes, of all things? Why not simply poison, or explosives? Why this extraordinary fascination for "bioterror," for an alleged threat of attack by pathogenic agents? The answer to this question is a bit complicated. It takes us back to the Prologue, back to the BBC documentary film in which chickenpox viruses in the shape of spiky flying objects swoosh around amid skyscrapers. "Once inside," intones the film's narrator, "they're ready to take over." The Arab airplane assassins seemed to fit this image. And then anthrax turned up in the city—portrayed in the film as a collective body.

As I have argued, the anthrax letters were so effective not just because the bacterium is a real pathogen but because the letters were associated with the September 11 attacks. The little pinch of anthrax brought to life a narrative that since the latter half of the 1990s had been repeatedly and ever more frequently invoked. But why did this bioterror story find such eager listeners? Why did people in the United States and in Europe believe it then and, obviously, still? What made the supposed connection between Al Qaeda's terror and the letters so convincing? How, in speaking of Iraq, was George Bush able to conjure up "poison, disease, and gas"?

The answer is simply this: *because the story is phantasmatic at its very core;* because, in other words, the bioterror story taps into a whole way of perceiving with deep historical roots and a pattern of prejudice characteristic of western imagination. I want to show that it is no coincidence that, following September 11, bacteria came to be associated with the "foreign" bodies of Arabs. In Bush's speeches, as well in the media, this phantasm appeared whenever the specter of future—obviously "Islamic"—terrorists was raised. As we have seen, in September 2001, Judith Miller spoke of "Muslim martyrs" infecting themselves with Ebola to disperse the lethal virus in American shopping malls, and in January 2003 the Austrian TV magazine "Report" reiterated this image: "All it takes to contaminate a city of thousands of residents with smallpox is for one infected suicide attacker to hang around a crowded location and cough." The Arab terrorist—the suicide attacker par excellence—is a living biobomb, his body a pathogenic agent. We need to understand that so-called threat analysis of this ilk derives its coherence from a deep-seated racist phantasm that at the same time must be repressed in the broader interest of multiculturalism.

To get these ideas across, I want to follow a trail created—as many others—by the signifier anthrax. As the founder of the heavy metal band said, "Anthrax" was just a name, and even the Anthrax band members played with the "cool" and "aggressive" sound of the name and tried out various meaning effects. Already years earlier, an element of meaning in the Anthrax name re-emerged from the depths of history that ostensibly had nothing to do with bioterror and the Iraq War but was circulated by the band as promo button: a caricature strongly reminiscent of that of the Jew in National Socialist propaganda. This in-your-face logo evolved over time. In January 2002 a three-dimensional face appeared on the band's website shaped as a spider's body. It most likely was not intended to reactivate an association between Jews and vermin, a point I will come back to later.[1] The face subsequently reappeared, without the spider body, in a variety of forms. But in any event Anthrax imbued the face—which became the band's brand—with new meaning. From 2002, apart from information about their music and band members, the Anthrax home page also displayed "Anthrax Disease Information" and "Daily updated news on anthrax outbreaks." In particular, in February 2003 the site featured the face of the band's Jew-like logo hiding a text one could reveal by running the cursor over the face: "We're Coming For Saddam!"

Anthrax (the band) had made the same sort of signifier shifts that we talked about in the first part of this essay: "anthrax" meant something again, and the author of the anthrax letters was named Saddam. Anthrax (the band) called for war. "Anthrax" meant war.

After the Iraq War, the band reupdated its website. The band's home page showed a newly returned GI from Operation Iraqi Freedom. The first thing he did when he got home was to buy the new Anthrax CD. "Seriously, I wanna thank you all for being creative and doing something with the freedoms you all have! I do feel like the military tends to get underappreciated, but in the same respect, I don't want to underappreciate the artists, poets, writers, actors, etc. I may defend our way of life, but you all make it worth living. I've been a fan forever, and just want to say thanx. Keep up the good work, and I look forward to more music from you all . . . derek."[2]

Jews, Arabs, Sudanese

It is no accident that among the store of available twentieth-century images the band Anthrax chose of all things the caricature of a thuggish, ill-spirited, and physically repulsive-looking Jew. Now, whether Anthrax were using the image to present themselves as cool, assertive foreign bodies or were simply being provocative by playing with the Nazi denunciation of Jews as "foreign bodies" is hard to say. At any rate, the perception of Jews as foreign bodies has the longer tradition in the western imagination. In European societies, "the" Jew has always served as the living representation of the stranger par excellence, akin to Arabs or blacks, whom one rarely encountered. Moreover, the image of the Jew was also linked with a particularly interesting narrative that is relevant to our context. Since antiquity, time and again Jews were accused of conspiring against people of different faiths and of spreading poison. Such rumors reached a crescendo during the Black Death in the first half of the fourteenth century. Both the general population and lay and church officeholders were convinced that Jews were inciting lepers to poison wells and kill off Christians. Lepers and, soon, also Jews were said to be the direct perpetrators of such crimes.

On the basis of these accusations, Jews in many cities and other locations in Europe were summarily burned. But the actual backers of this long-standing intrigue were thought to be Arabs. The "King of Granada" or the "Sultan of Babylon" presumably supplied the poison and the financing. Since Arabs could not move about Europe without attracting attention, they paid Jews large bribes in exchange for acting as secret but fairly well integrated middlemen to do the deed and to report their results in writing to the Sultan, as later attested by allegedly confiscated letters.[3] This persistent rumor—the mother of all conspiracy theories—cost the lives of innocent hundreds. And "Babylon" served as the symbol for a city that by the Middle Ages no longer existed but that played precisely the same role Baghdad, situated some 80 kilometers from the ruins of Babylon, does today. For remember, according to the Apocalypse of St. John, Babylon was the seat of all evil: "Babylon the great is fallen, is fallen, and is become the habitation of devils, and the hold of every foul spirit, and a cage of every unclean and hateful bird" (Revelation 18:2).

The medieval explanatory model of a plot to kill Christians in which Jews play a central role turns up again in a number of variations, even though the Arabs temporarily disappear from view. In 1831, when Europe was threatened by cholera for the first time, Silesia and Prussia did not aim their first-line preventive measures—border closings and import restrictions—only at Jews in Poland whose livelihood was trade.[4] In fact, rumors flew all across Central Europe from the Baltic Sea to Berlin, Krakow, Vienna, and Hungary that Jews had spread poison and contaminated wells; in Hungary, the result was pogroms.[5] With the cholera outbreak of the early 1830s, when for the first time the disease spread from the Indian subcontinent over trade routes, hitching a ride with troop movements to Russia and from there to western Europe, a particular pattern of perception was established for all epidemic threats.[6]

This perception included Jews but often encompassed a much more general enemy: those who advocated contagion theory and who believed that diseases of this kind basically originated in the "Orient" as "Indian angels of death" or

"Asiatic Hydra." (In contrast, many physicians of the middle nineteenth century believed in local, "sanitary" causes.)[7] The diseases were described as a mortal threat over which neither the military defenses arrayed in the *cordons sanitaires* still deployed in the 1830s nor the "soldiers" of the body had any dominion.[8] In 1831 a native of Danzig noted in regard to the cholera raging in the city: "There we sat, encircled by loaded rifles . . . the Asiatic evil and lascivious women in our midst."[9]

That infectious diseases were "Asiatic" at source seemed to brook no reasonable doubt, and more than a few observers drew connections between an "oriental" way of living and diseases such as plague and cholera. At the 37th meeting of German Scientists and Physicians in Berlin in 1862, a certain Dr. Theodor August Stamm remarked: "Intellectually neglected humans produce epidemics and aid their spread. Educational development is essential to the health of humans. The freer a country, the tidier—the tidier, the healthier . . . Just look at East India. So long as countries remain in the grip of despots and hypocrisy, they can never be healthy. Freedom of thought is harmless; however, epidemics of superstition produce bodily epidemics."[10]

A decade or two later, in the early era of bacteriology, when contagion theory held sway, the metaphorical excess and resulting anxiety in the discourse on infection were still going strong. Since the 1870s, germ theorists had begun to link the perception of bodily invasion by pathogenic microorganisms and the old explanatory model of a Germany invaded by carriers of bacteria from the East. What was coming into the body first had to penetrate Central Europe, for example, in the case of typhoid fever, from "Polish Jews" or "from Russia"; in the typhus epidemic of 1879–80 "from outside," from Silesian "homeless people," from the "heart of the Middle East," from "Hungarian mousetrap dealers," but also from "Russian students" studying in Switzerland.[11] The list of such examples goes on: in the late nineteenth century, germ theory held that diseases were caused by pathogenic microorganisms that could only come from outside.

In marginal notes to laboratory findings and in the dry epidemiological reports of disease outbreaks, this metaphorical personification of "outsideness" also crept into the medical literature. Bacteria are "like" Poles, Jews, Slavs, Russians. They come from the East, ultimately from Asia. So although the lethal designer virus in Tom Clancy's *Rainbow Six* was constructed in an American pharmaceutical laboratory, it is hardly surprising that its name evokes a threat from the East. For, according to the gene technologist who wanted to baptize her fiendish product in the laboratory, "You couldn't just call it anything, could you?" She did not have to ponder her choice for long: "Shiva, she thought. Yes, the most complex and interesting of the Hindu gods, by turns the Destroyer and the Restorer, who controlled poison meant to destroy mankind, and one of whose consorts was Kali, the goddess of death herself. Shiva. *Perfect.*"[12]

Since the early days of germ theory, then, descriptions of pathogenic microorganisms employed a language which not only gave foreign invaders a central role but which since the 1880s also mimicked the language of war or a social Darwinist battle for existence.[13] The renowned German biologist Ferdinand Cohn, one of Robert Koch's teachers and mentors, noted as early as 1872, in the manner of social Darwinists, that phagocytes "usually exterminate defeated cells at once."[14] But in this field it can be difficult to keep technical terms and metaphors apart and to sort out overlapping and contradictory notions of infectious processes. Robert Koch remained skeptical of the explicit social Darwinist imagery of a "battle for existence" to describe infection.[15] Yet both his germ science and his school were distinctly based on the belief that bacteria are "enemies" that are best destroyed outside of the body through decontamination. Others stressed the struggle between phagocytes and invading bacteria inside the body. Elie Metchnikoff, the most famous and perhaps the most influential of these bacteriologists, described in 1884 how phagocytes combat their "enemies" on the *champs de bataille,* the battlefield of infection.[16]

Toward the end of the nineteenth century, and after innumerable factional disputes over the interpretation of infectious processes, two opposing schools of thought had emerged in germ theory and the new field of immunology: the "cellularists," who moved the dispute to the foreground, and the "humoralists," who showed with increasing success how the "foreignness" of bacteria made them dangerous for the body in that they either produced or unleashed certain toxins.[17] In the opening years of the twentieth century, pioneering immunologists such as August Wassermann, Emil Behring, Paul Ehrlich, and Emile Roux had already established that the body's defense system also employed "chemical agents": antibodies that block the effects of bacterial toxins.[18]

Or, put another way, and in a metaphorical nutshell: For the (largely French) school of Metchnikoff, the body's response to pathogenic microorganisms was a battle of man against man, cells against microbes. In the German bacterial discourse initiated by Robert Koch, it took on more the flavor of chemical warfare, in which a more rapid response and the ability to release in short order huge amounts of combat agents tipped the scales toward life or death. The history of the conceptualization of this "battle" of the body against microorganisms is complicated and does not concern us here.[19] Instead, I quote—by way of synthesizing the arguments and positions of the time—the 1913 formulation of the English physiologist Sir Edward Schäfer, which evokes a futuristic image of total war: "The outcome of an illness depends therefore on the outcome of the fight between hostile forces: the microbes on the one hand, and the cells of the body on the other. Both fight with chemical weapons. If the body's cells fail to destroy the invading organisms, then the invaders end up destroying the tissue of the body—for this is a fight to the finish."[20]

Whereas these words seem to presage the thunder roll of the First World War, the metaphorical essence of classical germ theory was actually formulated in 1885. At that time, the great German physician and anthropologist Rudolf Virchow, himself no germ theorist but very familiar with the Koch school, cited approvingly an observation in the Paris *Journal médical quotidien* concerning the characterization of microbes—in those days, microbes caused a sensation among scientists and mesmerized laymen. Bacteria, the article stated, "are independent beings, infinitely small, but prolific, with racial features, living in diverse milieus, coming from outside, penetrating the organism like Sudanese, ravaging it through invasion and conquest, without regard for kinship or allegiance."[21]

The metaphor seemed to ring true—like any metaphor. In 1881, Muhammad Admad bin Abdallah, a mystic from Dongola in Sudan who claimed to be a Mahdi (successor to the prophet), called for revolt in the name of the true faith against Ottoman-Egyptian rule, which since 1822 had driven some two million Sudanese into slavery. Though still nominally Ottoman, the Egyptian vice royalty had for a while been under the influence first of France, then of Britain, and at the time was subjugating the Sudanese—considered to be especially courageous warriors—in order to recruit troops for its army. But after 1860, the viceroy began increasingly to enlist unemployed English and American officers to help end the increasingly intolerable pursuit of slaves being conducted by his own Turkish-trained mercenaries.

Thus, in 1874, Charles George Gordon, a Scot known as Gordon Pasha, was appointed governor-general of the entire Sudan to lead the government's business in Khartoum. Together with a band of European adventurers, ex-officers, mercenaries, and colonialists, Gordon attempted to assert vaguely defined European-Egyptian interests in the midst of local conflicts and guerilla skirmishes, to end slavery (the politically correct concern of the English colonial power was in the air, though Gordon failed), and at the same time to plunder whole regions. At any rate, the political event to which the metaphor in the *Journal médical quotidien* referred was a shock for the European reading public: The—not yet so-called—"Islamic" Sudanese Mahdi's troops were actually able to push back the Ottoman-Egyptian colonial troops and bring them to crushing defeat in bloody battles. In January 1885, after a ten-month siege, the Mahdis took the capital Khartoum. In the exhausting slaughter that followed, Gordon Pasha, too, was killed. Despite its belligerence, Sudanese Mahdi rule did not last long. In 1898 the Anglo-Egyptian army defeated the Sudanese, and as a result in 1899 Sudan became a British colony.[22]

The apparently obvious metaphor of the dangerous Sudanese conquering far-off Khartoum successfully hid the real colonial balance of power. Since 1830, the French had been waging wars of conquest from Algeria and West Africa to Senegal. By 1847 they had completely subdued Algeria, and in 1881 occupied Tunisia, which they made a protectorate in 1883. In 1884 Cameroon and Nigeria came into German and British possession, respectively. At the same time, mainly the British and French were dividing up between them the regions of West Africa between Nigeria and the Gambia. In 1884 German Southwest Africa took shape in the area that is now Namibia. The British occupied Egypt in 1882; in 1884 they conquered the vast eastern portion of the continent; and they even extended their zone of influence in South Africa.

The year 1885 was a very good one for fearing the "Sudanese." Whereas their motivation was to liberate Khartoum (their own Sudanese-Arabic culture and settlement) from colonial troops, the impetus behind the colonial conquests, invasions from outside, subjugation of foreign "organisms"—foreign bodies, foreign peoples, foreign regions—was the "right of invasion and conquest." That is what the European powers were really after in their frenetic "scramble for Africa." In using a convincing metaphoric image to convey the dangers of bacteria, the phantasmic reversal was complete. The metaphor was neither casual nor coincidental but, rather, consistent with a broad racist plot. The colonial wars of conquest in the nineteenth and early twentieth centuries were commonly wars of extermination.[23] Yet the real threat was perceived to be blacks, Arabs, and Asians. In Europe, this "peril" was naturally always invisible; only in ethnographic exhibitions *(Völkerschauen)* could the paying public see the dangerous "Sudanese" for themselves.

Parasites

The bacteriological certainty established toward the end of the nineteenth century that potentially pathogenic microorganisms were enemies that needed to be wiped out had very practical consequences. Beginning in 1907, the health authorities of the German Reich had been building their own sanitary barrier along the Polish border, that is, a chain of initially ten decontamination stations where emigrating Polish and Galician Jews were "deloused." The British historian Paul Weindling showed that these transit stations were intended, on the one hand, as protection against "Asiatic epidemics" and, on the other hand, helped to ensure in cooperation with port health authorities in Hamburg and Bremerhaven that emigrants to America would carry no "Asiatic" diseases with them to the New World. The German authorities were following a policy that had been decided on the insistence of the United States at the 1881 International Conference for Public

Health in Washington: immigration countries—especially the United States—had the right to check the health of incoming passengers right at the ports of entry. In other words, German hygienic measures for securing the borders to the east, for "delousing," and for rigorous health policing of poor Jewish emigrants at German seaports resulted from massive American pressure with antisemitic undertones.[24]

During the First World War, the "enemy" imagery of bacteriology in Germany was doubly reinforced in that Koch's students occupied all the major posts in the health authorities and the medical corps. Moreover, the real war against the enemies from the East seemed to confirm the bacteriological world view. As already mentioned, this world view increasingly included the more or less subliminal metaphorical equating of disease agents with their human carriers. In particular, Polish Jews were seen as the "natural" hosts of lice, which in 1914 were shown to be the animal vectors of typhoid fever. There was enough empirical evidence around to make this bacteriological pattern of perception seem quite "true." It was put into practice for the first time between 1914 and 1918, when German troops launched a radical sanitary policy conceived and directed by bacteriologists of the Koch school. The policy consisted in extensive and degrading disinfection campaigns against the Jewish people in occupied Poland and against Jewish, Polish, and Russian slave laborers and immigrants.[25]

But for the bacteriologists, this apogee meant the end of their classical epoch. For one thing, since the turn of the century, British epidemiologists and immunologists had already been developing perceptual patterns more complex than metaphors of "invasion" and "enemies"—concepts that emphasized a kind of equilibrium and coexistence with "foreign" microorganisms. The end of the First World War saw the collapse of the German lines of defense and the literal defeat of countless lice-infected German soldiers across the now no longer militarily nor epidemiologically secured eastern borders. Even within Germany the influence of many of the old germ models and images declined dramatically. Now the infectious were not Poles, Jews, or Slavs but the country's own conquered soldiers.[26] It was no longer possible to draw a clear line between outside and inside.

Parallel to that, new diseases appeared, for instance, the Spanish flu of 1918, which spread like wildfire and seemingly everywhere at once, and new epidemics such as meningitis or polio. Their pattern of diffusion no longer fit the familiar model of invasion and defense. They seemed to be diseases that came "not from outside, but from inside," as Andrew Mendelsohn writes. According to the thinking of epidemiologists in the early 1900s, the "new" epidemics arose "at home and, indeed, 'sporadically,' that is, without any obvious relationship to the source and spread of infection."[27] In 1921 the British Ministry of Health even devised a pithy, though short-lived, new metaphor for this concept of infectious disease: the flu was "essentially an internal problem for all countries; there is no question of barring the wolf from the sheepfold; it has for years been lying down with the lamb."[28]

I mention these things, which in the twentieth century had ramifications for epidemiology as a largely marginal medical science, because on the whole they represent a lost opportunity. Despite the convulsions of the First World War, the incidental findings of epidemiologists failed to challenge in any substantial way the basic concepts that the nascent field of immunology inherited from germ theory—that is, the theory of "defense" or of "exposure" to foreign microorganisms. In 1935 the Polish bacteriologist and epistemologist Ludwik Fleck remarked, "All of immunological science is saturated with primitive combat metaphors." The epidemiology of infectious disease was admittedly complex. But, as Fleck tartly noted, the familiar and alluring "invasion" and "enemy" imagery persisted as a primary explanatory model despite there being "not a shred of experimental evidence that would persuade an impartial observer to hold such a view."[29]

The situation is little changed today, even though there is no lack of proposals to abandon the semantics of "combat" and "defense" in favor of language that acknowledges the difference between self and non-self in a nonmilitaristic way.[30] Recently, Nobel prize winner Joshua Lederberg also advocated switching from warlike imagery to a more ecological model as the paradigm for describing the relationship between microorganisms and human hosts in bacteriological and immunological research.[31]

But for our purposes, the *popular* acceptance of twentieth-century bacteriological and immunological language is what matters. At this level, in particular, the allure of the simple "enemy" and "invasion" concept was so strong that epidemiological findings that should have made an impression did not, even today. Indeed, in the twentieth century the most aggressive form of the perception of the enemy as vermin—that is, those who identify the carriers of a disease with the agent itself—began to shape our view of the world and to guide action far beyond the scope of bacteriology as a science. Especially in Europe and in the case of typhoid fever—the most dangerous form of typhus—equating the microbe/louse with the human carrier had an effect that can hardly be overestimated.

In Germany, already in the concluding stages of the First World War, gas warfare technologies, bacteriological methods of disinfecting soldiers, clothes, and buildings, as well as the new science of entomology with its experience of pest control in agriculture and forestry came together in a new scientific field. In this context, beginning in 1922, epidemic and pest experts employed the hydrocyanic acid gas Zyklon B to exterminate parasites.[32] In the Second World War, these developments in fin-de-siècle germ imagery, the sanitary policies of the First World War, and the pesticide technologies of the period between the two wars reached their by no means inevitable, but—within the framework of this discourse nonetheless conceivable—deadly endpoint. The metaphorical link between bacteria and carriers of bacteria was analogous to the shift from the individual body to the *Volkskörper* (national body) and true to a germ imagery of "the enemy" that was taken literally by Hitler and many hundred thousands of other perpetrators, sympathizers, and bystanders and validated by popular science.[33] It culminated in the murder of millions of Jews and other so-called *Volksschädlinge* (human vermin) in "baths and inhalation rooms" and under the "decontamination showers" of the National Socialist extermination camps.[34]

The annihilation of millions of people on an industrial scale in the Nazi camps under the guise of "hygiene" and "bacterial cleansing" may have been a historical anomaly. But equating enemies with pests, parasites, or pathogens is not.[35] Much of the most powerful political discourse of the twentieth century was poisoned in a very real sense by metaphors that centered on the idea of the infected and infecting body and denounced "enemies" as "microbes" or "parasites." One little-known example in this context is the genocide of Armenians in Turkey in 1915–16. In trying to build a western-oriented secular state out of the rubble of the disintegrating Ottoman empire, the Young Turks eagerly absorbed European turn-of-the-century biologistic ideologies. These ideologies represented a modern alternative to the Islamic rhetoric that they felt got in the way of progress and that served more to keep up appearances. The leaders of the Young Turks declared baldly that they intended to "eradicate" the Armenian Christians as "microbes in the body of the fatherland"—and then proceeded to do it.[36]

But the revolutionary left also subscribed to the idea that political opponents or the enemies of the fatherland were "vermin" that should be wiped out. A famous Russian propaganda poster from 1917 shows Lenin with a broom in his hand ridding not just "Russian land" but the entire globe of so-called social parasites: "Comrade Lenin cleans the Earth of all the unclean."[37] According to historian Gerd Koenen, the early days of the Soviet Union saw the proliferation of a "regular demonology" of filth, a political propaganda in which the enemies were repeatedly stigmatized with biological and not infrequently bacteriological labels. Lenin spoke downright obsessively of vermin, of "parasites," "bedbugs," "scoundrel fleas" that deserved to be squashed, and finally of the necessary "cleansing" and the "extermination" of "harmful elements" that menaced the body of the new Russia.[38]

As rationale for the ubiquitous, man-eating terror that his administration would unleash in the following years, Felix Dzierzynski, head of the Tscheka secret police established in 1917, used words that sound oddly familiar today: "We are at war, and it is the cruelest of fronts, to be sure, because the enemy goes about masked, and it is a life-and-death battle." With a radicalism probably unlike any before in history, postrevolutionary Russia was "cleansed": "From top to bottom, layer after layer of the old structure of society was cleared away in this fiercely led civil war . . . At the same time, in many respects the Russian revolution represented a first, often brutal segregation of ethnicity and culture" that in sheer breadth and number of victims far surpassed the population policy of the Nazis.[39]

The utopia of cleanliness, the phantasm of purity, and the fear of contamination of the social body were pervasive features of all twentieth-century political ideologies. Sometimes they were no more than a shadow; at other times they functioned as a tactical strategy, for example, the depiction of "Louseous Japanica" in the U.S. Marines' magazine *Leatherneck* in 1945; for the most part, though, they formed its innermost core.[40] The phantasm of purity itself is no ideology but an *idée fixe* that in retrospect is perhaps better described as one of the basic command codes of political language in the twentieth century. It can spread throughout the symbolic system of a society, producing biologistic images on the screen of the political imagination. This code is a small bit of language, a short chain of signifiers, a sprinkle of semantics, a few simple rules for metaphorical shifts, and two or three application instructions, not more. It may tip a social system into authoritarianism and more often than not propel it into a war of aggression or to genocide of its own population.

6

Infection, the Metaphor of Globalization

The enemy is invisible, or at best as visible as vermin; it infests the uncorrupted body of the nation and must be eradicated. The idealistic language of liberalism, rights of man, democracy, and equality, handed down from the Enlightenment of the eighteenth century and the nation-state ethos of the nineteenth century, was no match for the twentieth-century's semantics of the national body, purity, pests, and social parasites. The Second World War was perhaps only a chapter, a most awful one, in a saga of struggles that may arguably be understood as a fundamental conflict between two basic political and symbolic patterns of perception and action. It pitted a totalitarian politics of purity—of bodies, races, and classes—against the acceptance of irreducible diversity. In this struggle, however, the lines were anything but clear.

Indeed, the Soviet Union, which during the Second World War was forced by Hitler into a pact with the Allies, carried out its worst political and social "cleansings" in the late 1930s. But the western democracies of the twentieth century, too, were enamored with the possibilities of quashing difference and otherness and using biologistic terminology to trim the edges of society. Eugenic "improvement" of the national body was no Nazi invention but a stratagem of modern society that from the outset had some support from the left. Nor was the idea of a "racially" homogeneous society limited in any way to Hitler's Germany. The fight against foreign infiltration was an integral part of the politics of western states in the twentieth century.

It is a commonplace to assert that although American culture remains a culture of immigrants with very diverse roots, it was nevertheless obviously shaped by the Puritanical legacy of the Pilgrim fathers as well as succeeding Protestant immigrants from central and northern Europe.[1] And just as in Europe, in the United States the discoveries of germ theory at the end of the nineteenth century led to a cultural change typical of the modern age. The historian Nancy Tomes has shown how germ phobia influenced American society in the late nineteenth and early twentieth centuries: from the new male fashion of shaving to demonstrate cleanliness, to the sanitation of households and redevelopment of cities, to the anti-tuberculosis campaign in the period between the wars and the fear of infectious disease being introduced by immigrants.[2]

I hesitate to say, however, which of these elements was the more important: Was it the country's fundamental openness to immigration, which did promote a historically unique mixture of people and ways of life—with the proviso, however, that immigrants abandon their identities to a "melting pot" and consciously and actively become "American"? Or were the germ phobia and hygiene campaign that Tomes analyzes the symptom of a phantasm of purity, rooted in Puritanism, that led not only to smooth-shaven men but to antisemitic tendencies during the 1930s and also found an echo in the rabid anticommunism of the 1950s and the sometimes violent opposition to racial integration of the 1960s? The McCarthyism and political language of the United States during the Cold War, with its pervasive fear of the invisible, infiltrating enemy (the "mole" of espionage fiction), was probably the clearest expression of these ideas of purity. And certainly—mediated by history and kept alive through the Republican administrations of Nixon and Reagan—they belong to the prehistory of the current thinking about threat in the era of Bush II.

Particularly striking in the case of McCarthyism, and especially vis-à-vis J. Edgar Hoover, then head of the FBI, was the use of language referring to infection, foreign bodies, and disease. Communists were described as a sort of vermin, or compared with germs; Hoover called communism a "virus" and a "disease that spreads epidemics." Democrat Adlai Stevenson characterized communism as "worse than cancer, tuberculosis, and heart disease combined." J. Thomas Parnell, chairman of the House Un-American Activities Committee (HUAC), demanded that America be made "as pure as possible."[3]

John F. Kennedy was no less avid an anticommunist than McCarthy. Yet his presidency heralded a new epoch in which America left aside its obsession with communism and the infiltration of "invisible enemies" to become the very model of an open, liberal, and increasingly pluralistic society. Setbacks and countermovements notwithstanding, this cultural revolution started a megatrend. Over the next twenty years, it spawned a hegemonic form of mainstream democracy with many regional variations and—at least officially—rejection of the totalitarianism of purity.[4] Western societies became more "colored" and more polymorphic than ever before. The practical result is that they are now all connected to the rest of the world by trade relations, tourism, demographic shifts, live TV images, and global popular culture—not to mention the Internet.

And as I will show, this background is indispensable for understanding that the fear of "bioterror" is just a symptom: a symptom of anxiety at this global mixing. I will show that already in the language employed after September 11, the attackers were perceived to be "parasites" or "vermin." And in the United States, this perception awakened a racist desire to eradicate the terrorists like vermin—the response of a sovereign nation, which, under a "state of exception" (self-declared global war on terror), rids itself of its enemies. I will show further that inasmuch as the metaphor of "infection" is bound up with the process of globalization, "bioterror" becomes the epitome of fear, and the global war on terror a strategy for fighting "plagues" in an ever more complex world.

Terrorist Vermin

After 9/11, a pattern of expression emerged in the global media that harkened back disturbingly to the old way of talking about germ and parasite science. A few hours after the crash of the first airplane into the World Trade Center, a survivor, traumatized and far from any anxiety to avoid a politically incorrect wording, said of the collapsed north tower in a CNN interview: "Whatever we have to do to eradicate the country or the world of this—of this vermin, I just hope Bush will do whatever is necessary to get rid of them."[5] Even though the semantics here is confused—one can hardly "eradicate the country or the world"—from this point on, the talk of "terrorist vermin" and their "eradication" does not let up.

A 9/11 memorial Web page provides a good example: "It must be recognized that these vermin (and I apologize to our friends the rodents and cockroaches of the world for degrading them so but there are no words in the English language to describe how pitiful these terrorists are) exist because we allow them to exist. And it is time to put an end to this. Yes, the vermin will fight back, just as insects and rats will fight back and flee when you fumigate a building. Sometimes you just have to hold your nose, put on your gloves and clean out the garage . . . So again, all people's of the world. Make your choice, now. Are you civilized or are you vermin?"[6]

In the same spirit, the columnist Dana Johnson created a Web page titled "Awakened Eagles": "Message to terrorist vermin everywhere: You have awakened the Eagle, and it is suddenly very hungry for vermin."[7] The warlike-patriotic site "Madtrimmer"—"politically incorrect and proud of it"—contains images of bomb attacks on targets in Afghanistan, including a sequence titled "Al Qaida Vermin 1–7" that shows an enemy vehicle being hit and exploded. It also shows a picture of two decapitated heads, apparently belonging to Arabs, with their penises stuffed in their lifeless, open mouths. A person named Bill Fisher writes on another 9/11 memorial page: "We are united as one Nation in our resolve to destroy the terrorist vermin of the world and the governments who sanction them," adding, for good measure: "Terrorist vermin and harboring governments: Know that your days of living among the decent, law-abiding, freedom-loving peoples of the world are surely numbered and the count down for your eradication has started!! You WILL be destroyed, utterly and totally. We, the people of the United States of America demand this."[8] In the *nationalreview online,* the columnist Deroy Murdock called for all Americans to contribute "to help pinpoint and pulverize the terrorist vermin."[9] Finally, at the end of 2001, in *WorldNetDaily*, former FBI agent and columnist Gary Aldrich asserted that the "FBI, the CIA, military intelligence, and law enforcement agencies . . . will do a good job of ridding our neighborhoods of these vermin."[10]

These are only a few of the approximately 307,000 hits that Google returns when you feed in the keywords *terrorist* and *vermin*. The combination *terrorist* and *parasite* gives a similar result (365,000 hits in May 2006). The fact that it is not only the extreme right that uses such language on the Internet is significant. In the U.N. plenary session of December 6, 2002, with the prospect of war in Iraq looming, American Ambassador Richard S. Williamson stated that the U.S. military operation Enduring Freedom in Afghanistan "gave the Afghan people the chance to rid themselves of a terrorist parasite."[11]

In the American online journal *Infectious Disease News*, this "parasite" had already been identified in December 2001 in a commentary from editorial board member Alan Tice as a "microbe" with increasingly dangerous antibacterial resistance. Reminiscent of the Defense Department's bomb war posters, the CIA and the FBI turn up here as "pharmaceutical firms." However, writes Tice, "the parallels between infection control and bioterrorism go further . . . The terrorists who have begun this war seem to have done so to protect their way of life and their beliefs. Western civilization has threatened it, and so, they attack. It seems almost a fight for survival, as it is for microbes, which are trying to fight off the antimicrobials we have thrown at them."[12] Tice, an epidemiologist, seems to know what he is talking about: the "terrorist" is the icon par excellence of the "invisible enemy" that invades our body from outside, to destroy it from the inside. It is a foreign species that must be countered with antigerm measures.

Of course, this explicit likening of terrorists to microbes occurs seldom—in contrast to terrorist vermin or parasite. And yet talk of invisible terrorists insinuates itself only too easily into the figures of speech arising out of bacteriology. Even President Bush availed himself of such expressions in his "War on Terror" speech before Congress on September 20, 2001: "But the only way to defeat terrorism as a threat to our way of life is to stop it, eliminate it, and destroy it where it grows."[13] The metaphor is a biological one. Terrorism "grows"—like cancers, like parasites, but also like germs in a petri dish. The wording is not definite, but one feels uncomfortable with the subtext. At any rate, the only way to fight them is to "eliminate" and to "destroy" them. In the "Axis of Evil" speech at the end of January 2002, this rhetoric became even sharper. Bush began by saying, "Thousands of dangerous killers, schooled in the methods of murder, often supported by outlaw regimes, are now spread throughout the world like ticking time bombs, set to go off without warning." Here again, the semantics are all but clear. We cannot take this wording as proof, but it does raise the suspicion: Underhand and most probably unconsciously, these "time bombs" are imagined as "invisible enemies" like vermin, like parasites, like cancer cells—like microbes.

The terrorist "underworld," said Bush, getting to the point, "operates in remote jungles and deserts, and hides in the centers of large cities." The attack on September 11 was, as the president had already stated on that fatal day, an attack on "the world," on "civilization." Terrorists do not belong to the world but live and operate in "deserts" and "jungles" inimical to "civilized" life and from there infiltrate large cities where they "hide." That is how they threaten the body of the city, and the nation. "We are," said Bush in this respect, "not immune from attack."[14] Hence his "hope" that "all nations will heed our call, and eliminate the terrorist parasites who threaten their countries and our own."[15] The war on terror is a war of extermination by the civilized "world" on the "underworld" of parasites.

One person in particular did heed the call: British Prime Minister Tony Blair. Speaking of the fight against terrorism before the U.S. Congress on July 17, 2003, Blair managed to work in mention of poverty, dictatorial regimes, and fundamentalist Islam. "In the combination of these afflictions a new and deadly virus has emerged. The virus is terrorism."[16] And on August 23, 2005, the *Washington Post* carried an article titled "Terrorism as Virus." Regarding the missing "clarity or consensus on whom or what we are up against," the article earnestly if naively suggests calling the enemy by its rightful name: "One promising new approach builds on the parallels often drawn between terrorism and a mutating virus or metastasizing cancer. Although Islamist militancy is clearly not a disease in the clinical sense, it does exhibit qualities of a social contagion; there is something undeniably appealing or 'infectious' to many about the ideas and beliefs that motivate terrorists and their many supporters. Analyzing the terrorist threat with an epidemiological framework would give focus and direction to our effort in three areas."[17] I will come back to this suggestion by the authors of the *Washington Post* article in the Epilogue.

Racism and Sovereign Power

But first it is worthwhile to delve a little more deeply into these "parallels often drawn between terrorism and a mutating virus" and the attendant hidden meanings. For example, in a prepared speech before the Senate Judiciary Committee on September 25, 2001, Attorney General John Ashcroft showed himself to be no-nonsense on the war against terrorism: "Today I call upon Congress to act to strengthen our ability to fight this evil wherever it exists, and to ensure that the line between the civil and the savage, so brightly drawn on September 11, is never crossed again."[18] That was plain language, not metaphorical, like the talk of terrorism "as" virus. September 11 starkly revealed the dividing line between the civilized and the "savage." Not between the West and the East, Christendom and Islam, whites and Arabs—rather, between civilized people and uncivilized ones.[19] If ever there was a colonial and racist pattern of discourse, it is this duality. The "savages" are those who live in the underworld, in deserts and jungles. They are parasites that threaten our bodies. They are those who must die that we may live; they are the antigens, against whose attacks we are not "immune."

Racism, as Michel Foucault states so concisely, is the distinction between "what must live, and what must die."[20] Foucault argued that racism not only fixates on skin color and shape of the skull but above all must be understood as the difference between "healthy" and "sick," "strong" and "degenerate," "worth living" and "not worth living." "The fact that the other dies does not mean simply that I live in the sense that his death guarantees my safety; the death of the other, the death of the bad race, of the inferior one (or the degenerate, or the abnormal) is something that will make life in general healthier: healthier and purer."[21]

Our concern is not this direct biological connotation of health and illness but rather the political understanding of "the difference between life and death" as Ashcroft said before the Senate Judiciary Committee. And not only: Ostensibly what is at stake if the terrorist other is not eradicated is the survival of the free world, of civilization, that is, America. The language of the war on terror, which contrasts "us" and "vermin," "civilized" people and "savages," means taking a stand on who must live and who must die. After September 11 Bush repeatedly stressed that people had to decide: "Either you are with us, or you are with the terrorists."[22]

The implication of this rhetoric is more or less clear, though under the Bush administration's official policy on multiculturalism hardly anyone would dare utter it: metaphorically as a parasite or vermin, or even totally unmetaphorically as a "savage," the terrorist other is simply excluded from the realm of what counts as human. To dismiss such language as sheer rhetoric is to turn a blind eye to history. Racism has always tended to deny the humanity of the objects of its hatred: either biologically, by classing nonwhite humans with apes, or culturally, by arbitrarily defining "civilization." The "savages" on the wrong side of this line of demarcation lived in fear for their lives.

Since the early nineteenth century, western societies have battled the "savages" and commonly killed off "savage" people. For example, during his South American expedition to Argentina in 1833, shortly after the Argentine government decided to rid the pampas of Indians, Charles Darwin witnessed manhunts for Indians and the horror of their mass murder. On meeting up with General Rosas' blood-smeared soldiers on the Colorado River, Darwin was deeply repelled. But he later conceded that such extermination campaigns were probably inevitable against inferior "races": "At some future period, not very distant as measured by centuries, the civilized races of man will almost certainly exterminate and replace the savage races throughout the world."[23]

I do not mean to say that this form of exterminating racism is at work again today. But the arbitrary distinction between "savage" and "civilized" can be explained only in the light of this sorry tradition of western thinking. Which brings up a paradox: the traditional way of talking about enemies as vermin, pests, parasites, or germs is usually aimed in a racist, biologistic way at a "people" *(Volk)*, an ethnic group, a "race" that as such threatens to contaminate one's own national body *(Volkskörper)*. In contrast, the designation "terrorist" does not really fit into this model of ethnicization or biologization. As a proxy for the Bush administration, Ashcroft did indeed speak of terrorists as "savages," but he explicitly refused to say whether that meant Arabs. Let us leave aside any Freudian speculation that this denial could have meant anything but its opposite—though it is possible—and simply consider the statement: It isn't Arabs but terrorists who are "savage."

This distinction, however superficial, makes perfect sense. In an age of globalization and in a multicultural society, mobilizing the broad societal consensus needed to construct enemies for purely political purposes—as opposed to just chasing after criminals—would be undermined by explicit ethnic overtones deriving from traditional, garden-variety racism.

Yet it is precisely this form of racism that underpins all warlike talk about terrorists. As Foucault shows, the issue is not a psychological one. Even taking into account the increased stridency that followed September 11, the setting apart and exclusion of the "other" who must die so we may live could not have happened without the familiar, gutter brand of racism. Neither Bush nor Ashcroft nor other members of the administration, to say nothing of the Web pages I have quoted, resorted to the measured language of law enforcement. Their words, which make sense only in the context of twentieth-century history, betray a racist desire eager to eradicate and destroy the enemy like a "savage"—like a "parasite." On certain websites, this desire climaxes in a sort of obscene pleasure *(jouissance)*, real enjoyment at the imagined annihilation of the "terrorist vermin"—Castrate the Arab! Cut his head off! . . . the Eagle is suddenly very hungry for vermin!

"Sovereign is he who decides on the state of exception," wrote Carl Schmitt.[24] The war on terror is such a state of exception imposed to varying degrees by the U.S. government on policy, freedom of speech, and the world itself in the wake of September 11. Killing off the terrorist vermin thus appears to be the sovereign decision of the American Eagle. But this sovereign power operates in a problematic area. Because for the moral conscience of the West, the racist itch simply to wipe out the "parasites" is intolerable and therefore illegitimate. It creates the paradoxical political problem of fulfilling it without being openly racist and offending the sensibilities of a multicultural society—a concern articulated even on politically "incorrect" websites.[25] But the fact that the sovereign power in the war on terror kills people without charging them and absent legal proceedings, without justification and out of sight of the public, is no secret.

One noted example is the killing of both of Saddam Hussein's sons, Udai and Qusai, at the end of July 2003. They were not captured but were shot on sight when GIs cornered them in a house. A columnist for the *Boston Globe* reported the memorable exchange on this event between a reporter and General Ricardo Sanchez, commander of the American ground troops in Iraq. The reporter asked about the reasons for killing the sons and added: "The Americans are specialists in surrounding places, keeping people in them, holding up for a week if necessary, to make them surrender. These guys only had, it appears, AK-47s and you had an immense amount of firepower. Surely the possibility of the immense amount of information they could have given coalition forces, not to mention the trials that they could have been put on for war crimes, held out a much greater possibility of victory for you if you could have surrounded that house and just sat there until they came out, even if they were prepared to keep shooting."

SANCHEZ: Sir, that is speculation.

REPORTER: No sir, it's an operational question. Surely you must have considered this more seriously than you suggested.

SANCHEZ: Yes, it was considered, and we chose the course of action that we took.

REPORTER: Why, sir?

SANCHEZ: Next slide—or next question please?[26]

The sovereign power of the field commander needs no justification. But this position becomes increasingly difficult when the vanquished, apprehended, and rounded up become prisoners: Is it permissible to kill a prisoner who, as terrorist vermin, has been denied legitimate status as a soldier? The imprisoned Taliban and members of Al Qaeda—and other Arabs—in Guantanamo and Abu Ghraib seem to fall outside the legal sphere precisely because they cannot simply be killed, as one kills vermin. Yet they are deprived of the most basic rights to criminal proceedings and even martial law.

They are *homines sacri* in Agamben's terms: people protected neither by secular nor divine justice, yet there is no religious or legal rationale for killing them. They are outlaws, reduced to bare life, and simply "there," both inside and outside.[27] Agamben argues that since the early days of Roman law, the sovereign power of a ruler begets *homo sacer,* who has recourse only to his "bare life." He *is* nothing but "bare life," and it is precisely in that fact that he represents the inverse of the ruler's sovereign power. Indeed, according to Agamben, it is this exclusion itself that constitutes the power of the community and its law: "In Western politics, bare life has the peculiar privilege of being that whose exclusion founds the city of men."[28]

The prisoners of Guantanamo and Abu Ghraib are therefore not only the symptom but the real, material result of a paradoxical politics that rediscovered its enemy in terrorists. Although America has vowed to fight the enemy to the death and to completely obliterate him, for the moment it has hold of him but it cannot kill him. The terrorist enemy at Guantanamo cannot even be threatened with capital punishment, since that would imply recognizing his fundamental right to a trial. So prisoners are kept like animals in cages and occasionally let out for torture, which consists less in gross physical injuries than in systematic harm to the prisoners' cultural, religious, and sexual identity. This torture aims primarily to strip prisoners of the last symbolic shred of personal identity and reduce them to no more than bare life urinating out of fright.

Tom Clancy has something to say about these complicated issues of sovereignty, bare life, and the right to that life. Recall that John Clark, commander of the antiterror unit Rainbow Six, got the go-ahead from CIA chief Ed Foley for him and his top-secret team to chase after four hijacked aircraft with an AWACS plane. The trail leads them into the Brazilian rainforest, where fifty-odd environmental activists and bioterrorists have set up their second base, including a runway. The place is a hell far from civilization, not unlike the deep dark place where Colonel Kurtz in *Apocalypse Now* established his lawless reign of terror. It doesn't frighten Clark; just the opposite: "It was so much like Vietnam, Clark thought, riding in a helicopter over solid treetops of green."[29]

Clancy observes repeatedly that in his later CIA and Special Forces assignments in South America, Clark carried out operations that were frankly criminal. Now, in designing the Brazilian operation, Clark argues that no court is competent to deal with the bioterrorists' monstrous plans to annihilate humankind. They are an exception that in turn necessitates exceptional combat measures. "I'm not sure the law anticipates anything this big, and I'm not sure this is something we ever want to put in a public courtroom, y'know?" A subordinate from the FBI counters, "We have two missing girls they probably murdered . . . and that's a crime, both federal and state, and, Jesus, this [bioterrorist] conspiracy . . . that's why we have laws, Mr. Clark." Even Ed Foley, head of the CIA, has doubts: "But we can't *murder* these people, John! They're U.S. citizens with *rights,* remember?" Whereupon Clark replies, "I'll try something creative."[30]

His plan—at least the part he told Foley—was to provoke an uprising in the jungle and thereby kill the fifty-odd terrorists, among them women, who, since they were biochemists, physicians, and computer technicians, would hardly be skilled at using weapons. "Foley frowned mightily, worried about what would happen if this ever made the media, but there was no particular reason that it should. The special-operations community kept all manner of secrets, many of which would look bad in the public media."

"John," he said finally.
"Yeah, Ed?"'

"Make sure you don't get caught."

"Never happened yet, Ed," Rainbow Six reminded him.

"Approved," said the Director of Central Intelligence, wondering how the hell he'd ever explain this one to the President of the United States.[31]

But the plan was more diabolical than Clark let on, which is also why I cite it here. First off, the highly trained men of the Rainbow Six team intended nothing less than to draw the terrorists out from their hiding place in order to kill a dozen or so of them with hi-tech weapons. The legal pretense would be self-defense "under fire," but the reality was ambush: "This was pure murder," one Rainbow fighter reflected. Clark instructed the remaining bioterrorists by radio to lay down their weapons, come out of the buildings, and assemble on the runway. There he told them to disrobe, while his team burned the clothes and blasted all the buildings. That was Clark's solution. At the end of the novel, the sovereign power of the Rainbow Six commander consists precisely in determining a state of exception and creating bare life. From their departing helicopter, as the team has a final glimpse of the group of naked bioterrorists abandoned to nature, Clark's son-in-law and Rainbow group leader Domingo Chavez asks: "'Maybe a week, Mr. C?' . . . A graduate of the U.S. Army's Ranger School, he didn't think that *he* could survive very long in this place. 'If they're lucky,' Rainbow Six replied."[32]

Agamben argues that in the modern age, producing bare life has become the most important form of biopolitics. That would appear to be a debatable thesis—biopolitics, as Michel Foucault defines it, is primarily the production and optimization of life and not its surrender to death.[33] But it is not my intent to get into a theoretical discussion here. Clancy considers both possibilities. The novel ends by tacitly approving both the production of bare life which cannot be killed under law as well as its surrender to death through the sovereign power of "Rainbow Six" Clark. By the same token, in an epilog, Clancy has Clark's son-in-law Domingo cast a glimpse on the photo of his baby that he keeps on the desk in the barracks, and the novel ends with these words: "His son would grow into the Brave New World, and his father would be one of those who tried to ensure that it would be a safe one—for him and all the other kids whose main tasks were learning to walk and talk."[34] Two models: The production of life in the armed nuclear family on the side of the sovereign power, and, at the other extreme—under a state of exception—Domingo and Clark's exclusion of bare life and abandonment of it to death in the jungle for the security, even the survival, of the world.

Immigration and the Biopolitics of Infected Bodies

The *homines sacri* at Guantanamo and in Abu Ghraib, reporters' unanswered questions and the cockiness of American field commanders (and their political leaders) at being able to define a state of exception are elements of a power that cherishes sovereignty—in the sense of Carl Schmitt—more today than even ten years ago. This sovereignty corresponds to a certain structure of political imagination (and thus probably also the structure of political power) that is beginning to form in this era of globalization. We are already acquainted with some of its features: terrorists and the many racist images of them, the "bioterror" conjured up by the anthrax letters, and so on. But these features blend with perceptions of infectious disease and migratory shifts, leading to a new threat scenario, whose metaphorical essence is *infection.*

Let us stick primarily with empirical evidence. The connection between terror, bioterror, and infection of the human body and the nation through infiltrating foreign bodies was first introduced in the *National Security Strategy of the United States of America* of September 2002. According to the strategy, efforts to protect the homeland from threat make a virtue of necessity, to wit: "Emergency management systems will be better able to cope not just with terrorism but with all hazards. Our medical system will be strengthened to manage not just bioterror, but all infectious diseases and mass-casuality dangers. Our border control will not just stop terrorists, but improve the efficient movement of legitimate traffic."[35] In this condensation, distinctions seem superfluous and politics overlaps with epidemiology and epidemic control: terrorism, infectious disease, bioterror, and stemming illegal border traffic—the opposite of "legitimate traffic," in other words, clandestine immigration over the Mexican border. These separate issues conflate in a single brief paragraph to an "opportunity" for intensified control of infection at the borders of the national body. That the reality often looks quite different, and that even trivial budget bottlenecks hinder the ability of U.S. police personnel at the Mexican border to meet these ideological specifications, in no way alters the *idée fixe* of a national body that must be protected against "infection."

I do not mean to say that all immigration controls and restrictions are inspired by this idea—there are also sound economic and social reasons for immigration restrictions—nor do I mean to say that these restrictions are wrong in every case. My point is solely that there is a phantasm of a national "body" that requires protection from "infection." That is, phantasmagorical excess in the perception of reality fosters a hysterical and dangerous defense that threatens the liberal tenets of our society (as expressed, for example, in Samuel Huntington's idea of the threat to the "white identity" of the United States posed by Latino immigration).[36]

There are many examples of the specific fear born of the connection between infection and immigration. In May 2005 an article on the website of the Americans for Legal Immigration announced, "Illegals bring new disease outbreaks! Who's cooking your food?"[37] In February 2005 the conservative *Washington Times* reported with alarm: "Contagious diseases are entering the United States because of immigrants, illegal aliens, refugees, and travelers."[38] In 2003 sensational reports appeared concerning a rare blood disease—Chagas, a "parasitic infection common in Latin America"—that "is increasingly affecting the United States blood supply." However, the *New York Times* subsequently corrected its data to read that in 1996, in Los Angeles, only "0.010 percent" of donated blood showed antibodies to Chagas, "0.014 percent in 1997 and 0.018 percent in 1998," and observed that no one seemed to be much worried about a test for the disease.[39]

In the same year, in an article titled "Immigration and Disease: It's Enough to Make You Sick," the Phoenix *American Daily* volunteered its two cents on the world situation of microbiology: "The U.S. is not alone in its fight against illegal immigration and disease. Illegal Chinese immigrants to Europe are bringing to that continent malaria. Even Thailand has problems with diseases brought in with immigrants from Burma. The biological clock is ticking and a world health crisis is looming. Because the U.S. suffers more immigration than any country in the world, our health problems are growing faster. Certainly, the potential for biological disaster has come to the notice of terrorists and Al Qaeda."[40] Here, epidemiological problems, which in fact can be handled without difficulty and therefore are no cause for excitement, somehow become time bombs for a worldwide crisis—and, naturally, Al Qaeda cannot be far away.

There is no denying the real health dangers connected to movement of people worldwide; for me, the question is purely one of showing how quickly fact and fiction meld in this area, that is, how quickly medical terms can become political metaphors. In the BBC documentary cited in the Prologue, there is a scene with two border police who are peering through the dome of their low-flying helicopter, using a searchlight to scan through the dark to the dusty soil beneath them. Every night, says the narrator, hundreds of police on the Texas border with Mexico battle the most dangerous enemy that has ever threatened the United States from the south. They hunt for the countless illegal immigrants who scale the border fence to find work in America. But the next image, of a captive in an examination room, illustrates what the film "really" wants to say: the enemy is tuberculosis, which many of these immigrants have, and fighting it is one of the tasks for which these gruff-talking, broad-shouldered policemen are paid. On the X-ray screen in the U.S. Public Health Service mandatory isolation facility, the devastation already caused by this enemy in the lung of a gaunt Mexican is visible: an image like a map, showing the occupied terrority.[41]

Such images, which are always both metaphorical and more or less realistic—tuberculosis *is* prevalent among Mexican immigrants—appear to have accumulated in recent years. Particularly striking was a Reuters photo dated January 4, 2002, that shows a Kurdish refugee being helped out of a boat seized by Italian health officials near Bari. This Kurd is not just poor and politically persecuted, nor did he just enter Italy illegally—he is above all potentially infectious. Images such as this one are so unsettling because there are in fact medical reasons for handling refugees from eastern Turkey (where anthrax is endemic) with rubber gloves and protective masks. Yet the metaphorical excess in this image goes far beyond such realities. In the context of European history, it is hard to avoid equating the Kurd in this photograph with risk of infection.

Even more disturbing are images that show the torturers in Abu Ghraib wearing rubber gloves to reduce the chance of infection through their "contact" with Iraqis. Interpret them as you will: as evidence for the phantasm of the Iraqi other as a "germ"—or "simply" as the same sort of medical precaution a dentist might take. In either case, the torture victims give the impression of being potentially infectious, and thus dangerous. It is no accident that a right-wing Canadian website with the evocative name *immi-kill*—clearly equating one with the other—begins with a short, innocuous animated movie about handwashing.[42] Contact with immigrants makes you dirty. So although the protective masks in the photo of the Kurdish refugee and the rubber gloves of the American soldiers in Abu

Ghraib at first blush seem medically "rational," they invoke the racist phantasm that these aliens—Kurds, Iraqis—are infectious.

For some time now, the assumption that foreigners are "contagious" has also played a role in European Union discussions on immigration policy. Thus, for instance, a 1997 proposal passed unanimously by the plenary session of European Academies of Medicine demands strict immigration laws with explicit reference to "risks presented by immigration, and especially clandestine immigration, by bringing and propagating certain diseases, the most serious of which are tuberculosis, hepatitis B and C, and sexually transmitted diseases."[43] Illegal immigrants are the most infectious. The statements show how hard it is to separate "fiction" and "reality," phantasm and fact. There is clearly an intensified exchange and global spread of pathogenic microorganisms along the path of migration. Yet, isn't this exchange just as pronounced, or even more so, in worldwide tourist traffic? Can we really be certain that immigrants have so many more infectious diseases as to justify our fear of them, whereas business travelers, sex tourists, students, and all the other millions who cross international borders every day are somehow less affected by them? SARS has taught us differently.

Linking infection with foreigners is an appealing, popular, and long-established explanatory model. It is easy to invoke tangible evidence for it. Even easier, though not more plausible by any means, is to hold the least-desired arrivals—illegal migrant workers—responsible for infectious disease. But the *National Security Strategy* goes a significant step further. Here, illegal immigration and infectious disease appear *on the same level* as terror and bioterror. They are interchangeable forms of the threat to the national body, and in accordance with the model of germ invasion must be fought using the very same methods as political and military epidemiology and antiterror warfare. In this biologistic picture of politics, the "other" is not black or Arab or Asian but simply "the invader." The invader may appear in one of two forms: either the infected and infecting immigrant or a terrorist, always inclined to "bioterror"—that is, in contrast to the immigrant spreading the infection consciously and maliciously.

Thus we begin to see how terror turns into "bioterror." Terror is tantamount to the deadliest form of an infection that begins with streams of migrant workers swimming with individuals who "penetrate" the national body, carrying deadly germs with them—infectious diseases or superbugs. "Reality" and "fiction," conceptual and metaphorical discourse combine. Terror and "bioterror" have always meant precisely this "penetration" of the national body. Hence the signifier "bioterror" turns out to be the quintessential expression of fear of "infection" in the age of globalization. Whereas illegal migrant workers represent the average risk of infection, which every tourist is also exposed to, "bioterror" connotes the—apparently objective—maximal threat of pathogenic agents, against which there is no protection. The terrorist who brandishes such agents is an enemy to the death.

War of the Worlds

How does the global war on terror fit in this context? Global war evokes an ultimate battle between order and disorder, between civilization and savagery, between purity and infection—a war of the worlds. Such a war is the opposite of *infection,* mingling, migration, impurity, contamination—and bioterror. That becomes very clear in a film that, by way of emphasis, metaphorically extends the themes of war and threat of infection into the galaxy like a booming reverberation of the present global war on terror. The film is Steven Spielberg's *War of the Worlds,* released in the summer of 2005. I will show to what extent this film, as well as the war on terror, is concerned with biopolitics, that is, with infection, with the survival of "us," and with the death of aliens—of infectious invaders.

Spielberg's blockbuster might have benefited from a better leading actor than Tom Cruise, and also a more competent scriptwriter. But in any event, in playing around with the "future," films of this genre reflect reality, that is, the imaginable as a metaphor for the present; so, too, Spielberg's story about the invasion of Martians. The literary forerunner of the film is H. G. Wells's novel of 1898. The novelty of Spielberg's free adaptation of this ultimate threat scenario, and thus its political novelty, lies in the fact that Spielberg's aliens—not entirely finished gestating—have been hiding underground in the form of enormous, three-legged metal shells. It is in these shells that, one day, at the beginning of the twenty-first century, the aliens drive up out of the cracked asphalt like unleashed furies. The fragile bodies, and particularly the brains, of the Martians were transported only shortly before the attack by means of huge bolts of lightning (or pulses of energy?) into the machines hidden in the earth. But the point is that the lethal machines had been there for a long time, waiting to be mobilized: buried, secret—and right beneath us. In the period following 9/11, the alien is not just frightful but a killer monster, a combination of organism and machine, just like the airplanes that turned into suicidal killing machines. The alien of today is above all a sleeper. He needs only the impulse from outside to attack.

In one curious but important detail, Spielberg stays true to Wells. Toward the end of the film, it is obvious that the aliens are in trouble and no longer invulnerable. In the final scenes, they go down like dominoes, and the enormous shells perish with them. Why? Wells describes the dead Martians and their silent machines: "Laid in a row, were the Martians—*dead!*—slain by the putrefactive and disease bacteria against which their systems were unprepared."[44] This explanation is provided by an off-screen narrator as we are shown images of a dewdrop on a bud, which, after all the scenes of devastation, suggest life and blooming health. The camera then tracks into the dewdrop, revealing shots of microbes and, at the conclusion, even the threads of their DNA (images that also appear at the beginning of the film). They are the self-same microbes that did the aliens in: Thus, the aliens were not immune against these microbes. Wells comments on this in a way I also remember hearing in the film: "These germs of disease have taken toll of humanity since the beginning of things—taken toll of our prehuman ancestors since life began here.

But by virtue of this natural selection of our kind we have developed resisting power; to no germs do we succumb without a struggle, and to many—those that cause putrefaction in dead matter, for instance—our living frames are altogether immune . . . By the toll of a billion deaths man has bought his birthright of the earth, and it is his against all comers; it would still be his were the Martians ten times as mighty as they are. For neither do men live nor die in vain."[45]

Life and survival, victory over the Martins, the right of the firstborn—all at the simple "toll of a billion deaths . . . " The parallels between the film and the novel are intriguing: To what extent and in what ways does this film say something about our present? Was it by coincidence that it updates a novel that was written in the light of the arming of the German Kaiserreich and the nascent demise of the British empire, and that seems to have anticipated mechanized killing, the battle for survival of entire peoples and the millions killed in the First World War, and even gas warfare? That seems hardly likely.

Wells was referring (with an optimism characteristic of the time) to the apparently innocuous fact that through selection and adaptation, the human species ultimately succeeded in safeguarding life in a long, casualty-rich, but also largely foregone battle for survival against germs. Yet on any level and on any scale, the battle that is our subject here, and as it is portrayed in the stories by Wells and Spielberg—germs, "worlds"—is anything but innocuous. In explaining the setting of the story in the introductory paragraphs to his book, Wells writes that the Martians suffered from the cooling of their planet and looked around for a new habitat: "Their world is far gone in its cooling and this world is still crowded with life, but crowded only with what they regard as inferior animals. To carry warfare sunwards"—that is, toward Earth—"is, indeed, their only escape from the destruction that, generation after generation, creeps upon them." Then comes the key passage: "And before we judge of them too harshly we must remember what ruthless and utter destruction our own species has wrought, not only upon animals, such as the vanished bison and the dodo, but upon its inferior races. The Tasmanians, in spite of their human likeness, were entirely swept out of existence in a war of extermination waged by European immigrants, in the space of fifty years. Are we such apostles of mercy as to complain if the Martians warred in the same spirit?"[46]

That is what the *War of the Worlds* is about. Or rather, that is the context that gave rise to this story and this specific discourse, independent of whether the socialist Wells approves of it or not. For Wells, the annihilation of the "inferior races" such as the Tasmanians was cruel and ruthless, but all the same it was as inevitable as it was justifiable: "The intellectual side of man already admits," Wells wrote apropos, "that life is an incessant struggle for existence."[47] In his imaginary yet metaphorical representation of this battle, Wells turned the real situation on its head, allowing the invaders to die of infectious disease and not, as happened with all colonized or exterminated peoples, the victims of invasion. It makes clear that bacteriology is a science of winners: it is the knowledge of those who survived the battle of adaptation and whose superior immunity in the struggle for existence threatens the foreigners, the "racially inferior" aliens, with death by infection.

In this sense, the stories by Wells and Spielberg also have a social Darwinist and biopolitical context. The war of the "worlds" for survival begins the story, whereas the bacterial infection ends it and gives it its intrinsic "human" meaning. "We" have triumphed over other organisms, and thus "we" will also triumph over our enemies, who, like all living creatures, carry the battle for existence "sunwards." We will prevail because it is our birthright.

The Plague Model of Power

One could probably defend the proposition that the Iraq War represents the prelude to forthcoming battles over who will get the last reserves of fossil fuels—a battle "sunwards," as it were, a battle for the ultimate sources of energy: biopolitics and the struggle for survival on a global scale. But here I would like to pose another question: the question of the societal power at play that leads to such wars. In other words, the disposition ("dispositive," as Foucault puts it) of biopolitical power in the age of "infection." If it is true that in today's globalized world, "infection" has become the central metaphor, then from the perspective of power the metaphor implies mixing of people, new social and ideological complexities, and new asymmetric forms of warfare. On the other hand, globalization has also spawned a culture of infection that is as driven by the "viral" character of the Internet as it is by the uncontrollable, networklike structure of the rebellious "multitudes," as Michael Hardt and Antonio Negri have written. This fascination for subversive structures that defy hierarchy may go so far as to welcome mixing—the contamination, if you will—with alien life substances; or to radically (through art) celebrate the deadly, infectious threat as a way of infiltrating power and the established order.[48]

To understand these political dimensions of fear of—and conversely desire for—"infection," we first have to return to Foucault. *Discipline and Punish* (1975) raises the question, How did state authorities manage to extend discipline to behavior beyond that governed by prisons or the army? That is, How did emerging civil societies with their sense of personal liberty come to enforce discipline and to compel disciplined behavior among their citizens? And what made these disciplinary measures necessary? Foucault's answer at that time is interesting, even though we now know that he later revised it. The origin of modern law and order in the seventeenth century is the quarantine as a means used by urban authorities to defend against the plague. Plague regulations, which Foucault quotes, outlined a system of total control of borders and transit, all movements in the city, and forcibly locking people inside their homes. "It is a segmented, immobile, frozen space. Each individual is fixed in his place. And, if he moves, he does so at the risk of his life, contagion or punishment."[49]

Foucault sees in these regulations the bare bones of a

model: "This enclosed, segmented space, observed at every point, in which . . . each individual is constantly located, examined and distributed among the living beings, the sick and the dead—all this constitutes a compact model of the disciplinary mechanism. The plague is met by order, its function is to sort out every possible confusion: that of the disease, which is transmitted when bodies are mixed together." Thus the plague itself is a model—a counterprinciple for order, a "festival" of disorder as painted in the literature of seventeenth-century low-brow comedy; a festival of "suspended laws, lifted prohibitions, the frenzy of passing time, bodies mingling together without respect, individuals unmasked, abandoning their statutory identity and the figure under which they had been recognized, allowing a quite different truth to appear." To this "dream" of the plague as the breakdown of order, the authorities opposed a "political dream" for discipline, that is, "the penetration of regulation into even the smallest details of everyday life through the mediation of the complete hierarchy that assured the capillary functioning of power."[50]

Of course, Foucault is not speaking of cities where the plague really did break out but of the "utopia of the perfectly governed city," for which "the plague (envisaged as a possibility at least) is the trial in the course of which one may define ideally the exercise of disciplinary power . . . In order to see perfect disciplines functioning, rulers dreamt of the state of plague," just as "jurists place themselves in imagination in the state of nature." To dream of the state of plague is to perceive the dangerous limits of power: "Behind the disciplinary mechanisms can be read the haunting memory of 'contagions,' of the plague, of rebellions, crimes, vagabondage, desertions, people who appear and disappear, live and die in disorder."[51]

Foucault sees the perfected model of this form of government not in the city beset by plague but in Panopticon, designed by the legal philosopher Jeremy Bentham in 1787 as an architectural structure for all supervisory institutions: An annular building with solitary cells unconnected to each other but all directly visible from a central tower in the middle, which requires only a single supervisor to communicate to all individuals the well-justified impression of being continuously monitored. The advantages of this setup are obvious in Foucault's account of Bentham's text: "If the inmates are convicts, there is no danger of a plot . . . ; if they are patients, there is no danger of contagion; if they are madmen there is no risk of their committing violence upon one another; if they are schoolchildren, there is no copying, no noise, no chatter, no waste of time; if they are workers, there is no disorder, no theft, no coalition, none of those distractions that slow down the rate of work . . . The crowd, a compact mass, a locus of multiple exchanges, individualities merging together, a collective effect, is abolished and replaced by a collection of separated individualities."[52] The Panopticon is based on an elaborate system of visibility and in addition requires only the supervisor in the center as functionary, who in turn can be overseen and supplanted by any member of society—a democratic, not totalitarian conception of power.

Bentham's Panopticon has a critical task: to hinder contact of the inmates with one another so as to prevent communication, association, and infection. For Bentham, as in the *National Security Strategy,* movements of individuals not controlled by authorities are precisely analogous to movements of infectious substances. Within the blind spot of power, these substances seek out horizontal contacts and avoid going through the center. Foucault's thought model describes a power that both dreams of the plague and dreads infection; its solution is to separate bodies and to classify them in the solitary cells of the Panopticon. There is no question that real epidemics occasionally require short periods of quarantine and rigorous monitoring, and these old formulas proved their need and effectiveness once again during the SARS epidemic of 2003. But, as I have said before, Foucault is not concerned with actual infectious diseases but with a dreamed state of plague, an "anticipated" epidemic: in short, a model of infection whose function is to show what disorder can bring about and how power can counter it.

In that sense, the age of "bioterror" differs little from the seventeenth and eighteenth centuries. The biopolitics of infected bodies begins with the control of migrant workers. What better way to track down not just the infectious but also terrorists with viral killers in their hand baggage? Since the chances are good of obtaining a broad social consensus for the control of immigration along with defense against infectious diseases, terrorists, and superbugs, it follows that the chances are also good that this strategy will promote fear, discipline, and control. "Bioterror" is a code name for disorder, and people who invoke it dream of a power that ensures order.

And yet the plague model in its Benthamian form does not really seem pertinent to today's society. Postmodern societies tend no longer to be disciplinary societies. They are power structures that are based increasingly on the principle of flexible self-rule, in which each individual sees himself and everyone else simultaneously and can observe and also still be observed by all the others without having to be locked up in a "cell."[53] Does such a society need the regulatory hand of the state, which subjects individuals to the discipline, the *authority,* of an all-seeing eye? In particular, does a postmodern society really lust after epidemics? Must we have the phantasm of infection and of bioterror (translated into fear and epidemic control measures) to maintain order?

I do not want to launch into a theoretical discussion but simply to advance a little empirical evidence that the model of the dreamed plague, in modified form, is still at work today, maybe more than ever. We will see in the Epilogue that in his lecture on the "History of Governmentality" Foucault develops yet another model (the smallpox model), which conceives of an alternative style of dealing with infectious diseases and thereby also another—namely, liberal—form of power. But the point here is that power is once again dreaming of "plague."

America in particular has adapted the rigorous, central control of the Panopticon model, only adding to it mellower, more flexible forms of supervision. The new *Citizens' Preparedness Guide,* published in January 2002 by the U.S. Department of Justice, the National Crime Prevention Council, and the Freedom Corps under the title *United for a Stronger America,* projects the image of a harmonious socicty in which individuals mutually control themselves and each other. In this way control develops in accordance with a standard that is not—as in the old disciplinary societies—stipulated by an authority. Rather, it is supple, that is, contingent, dependent on and modifiable by the preferences of the individuals themselves. This society, as the guide describes it, consists of citizens, families, neighborhoods, and communities in which everybody knows everybody. The very first rule of terror defense under these circumstances is the following: "Know the routines. Be alert as you go about your daily business. This will help you to learn the normal routines of your neighborhood, community, and workplace. Understanding these routines will help you to spot anything out of place . . . Be on the lookout for suspicious activities such as unusual conduct in your neighborhood."[54]

Citizens are not only advised to stock up on groceries and water for five days but also to equip themselves with emergency power generators and to learn how to identify suspicious letters and to protect computers from viruses. Most important, they are repeatedly instructed how to observe neighbors, how to cooperate with them in observing other neighbors, and how to prepare together for the coming terrorist attack: "Just remember, it's your job to watch out and report."[55] The *Citizens' Preparedness Guide* contains some disquieting performatives. In reading it, you feel under the shadow of an unknown and unfathomable menace designated only as "terrorism"—a nameless foreign thing that reveals its behavior only outside of the normal and the routine. Even if you dismiss the guide as generally inconsequential, it still illustrates the tangible effects of superbug fear linked to the fear of ordinary terrorism. The guide is a manual for fear.[56]

America dreams of plague. The social system produced by this bioterror nightmare plays on two registers. On the one hand, in the sense of the plague mandate of the seventeenth century, state control is strengthened, borders more closely guarded, and citizens are confronted daily with a police presence that is without precedent in American history. Increased surveillance and rights of access of law enforcement agencies, enabled by the Patriot Act, prove that in a pinch even a postmodern society will fall back on a highly sophisticated center of power.[57] On the other hand, the *Citizens' Preparedness Guide* is the document of a society that is called to self-control, a bottom-up model of surveillance whose standards are flexible. It conducts reciprocal observation of routines and everyday occurrences as a civic duty, whatever these routines may be, whether the neighborhoods are white, black, Asian, or Arab. It is thus a double safety package, like a toolkit for the "city of the sun," for a strictly supervised multicultural utopia, which would be both a police state and a self-governed civil society.

But it hardly helps to be too cynical or pessimistic. In the Epilogue I will return to Foucault's idea of a liberal model of power with respect to infectious disease. Current critical thinking might do well to consider this model as a way of countering the dangerously simplified perceptions of the global war on terror.

Playing with Infection

America's dream of the plague is also a dream of globalization—a nightmare about dangerously open borders, streaming migrants, and uncontrolled exchange and contact. Even Clinton saw "bioterror" as the "dark side of globalization," which we now understand only too well. Globalization is the name for a barely contained worldwide "infectiousness" against which no Panopticon is any help, because it would have to lock up all of mankind. Nor is postmodern self-government much of a solution because many of the planet's cultures are regretfully still somewhat inflexible. The global war on terror reveals how obsessed the present U.S. administration is with the idea that America must root out every last agent of evil to be able to live freely in a globalized world. Anyone who dreams of plague has lost his peace of mind, and makes war the basis for policy.

As I have said, this nightmare has effects that may appear advantageous to power. But, as a complement to all those seventeenth-century schemes for order, was there not a tradition of low-brow comedy that, as Foucault writes, envisioned plague as a festival of disorder? Indeed, have there not always been manifold dreams of contagion? Don't some people dream a subversive dream of globalization in which infection is a fixture, meaning, perhaps, association, contact, and communication? Do they not express a tantalizing desire for wide-open borders and mixing of bodies—the unrestrained possibility of infection through the organic matter of an alien life itself? And what of the vicious appetite for infection, a theatrical game of decay, an evocation of death as the ultimate act of defiance? A comic literature of infection?

In America, the simultaneous fear of infection and secret longing to infect was associated at the beginning of the twentieth century with Mary Mallon (1869–1930). An Irish immigrant, she became known as one of the earliest cases of a healthy carrier of bacteria who was said to have maliciously desired to infect with typhus the respectable WASP families she cooked for. Marouf Hasian and Priscilla Wald have shown that the image of "Typhoid Mary" as the archetypal "infectious woman" was a sexist and racist construction.[58] Nonetheless, up to today Typhoid Mary has remained a rhetorical device that not only evokes the threat of infection but that makes a game of the evil desire to infect and to be infected.[59] There are countless underground bands called Typhoid or Typhoid Mary, and even more musicians who have sung about her. Take, for example, the Cheerleaders of the Apocalypse. The chorus of their song "Typhoid Mary" of 2000 makes Mary the heroine of a radical subversion, which topples a society already long infected:

> Typhoid Mary keep on burning
> Never will the river be the same
> Typhoid Mary keep on burning
> Girl you know you're not the one to blame
> Typhoid Mary there's no denying
> You spread the deadly fever and disease

> Typhoid Mary keep on shining
> Bring society down to its knees.[60]

With other bands, the desire for infection and the play with death is cruder and more brutal. For example, in 1999 the hardcore band Blood came out with the grim song "Ebola":

> something loudless, invisible has infected your body
> can't recognized that your days are counted
> . . .
> the virus got ya, the virus kills ya . . .
> you'll die a painful death and all
> who be with you get infected
> no way to escape
> the virus will survive![61]

Here the evil virus is the hero and the only survivor; the putrefied bodies—the putrefied society?—will die. But the subversion does not have to be so obvious. In other song lyrics, the foreign body (the virus) is a metaphor, an icon of one's own marginality; it signals one's own place in society. For instance, in their album *The Dangerous Doctrine of Empathy*, released in 2003, Thought Riot explores the image of punk as a dangerous microbe:

Here we come . . .
We're the poison, your new disease.
Soon you'll be nothing but a cryptic memory . . .
It's spreading, it's spreading,
We're spreading . . . here we come.[62]

Similarly, in response to an interview question about the title of their CD *New Era Viral Order,* Thee Maldoror Kollective responded: "We cannot recognize contagion cause we are infected, we cannot see the virus cause we're the viruses. And the antidote."[63] The underground is infected; the underground is viral.

Long before the bioterror hype, Alice Cooper, the old master of rock black humor, was penning lines that may well be the best expression of the subversive reversal of the fear of infection. His song "Nuclear Infected" came out of the reactor meltdown at Three Mile Island:

I'm nuclear infected
I really don't mind
I just go out when the sun goes down
And have a real good time.
. . .
I'm nuclear infected

Really ain't that bad
In fact it's about the best time
I guess I ever had
. . .
When I'm happy I glow yellow
When I'm sad I glow blue
And I glow red hot when I'm in bed with you.[64]

These last lines themselves read like an ironical commentary, before the fact, on the Department of Homeland Security's color-coded threat-level system—which became the target of endless derision. For example, at the Venice Biennale of 2005, the feminist art and action theater group Guerilla Girls exhibited a caustic critique of Bush in the form a huge poster in the well-known colors titled "The U.S. Homeland Terror Alert System for Women":

Low: President rides around on horse,
clears brush on ranch
Guarded: President appoints man to federal drug
administration who believes prayer is the best
treatment for pms

Elevated: President's economic policies result in largest job losses for women in 40 years

High: President declares abstinence his favorite form of birth control and the answer to AIDS epidemic

Severe: President claims women do have rights: they can join the army, fight unprovoked war, kill innocent people.[65]

Another, rather clever example chosen at random from the Internet (and abridged):

Low: Have ready access to long pants, cigarettes and aspirin . . .

Guarded: Oh boy! Family road trip to the U.S. Army Desert Chemical Depot in Tooele, Utah! . . .

Elevated: Tape pant legs closed at the ankle, shirts closed at the sleeves . . .

High: Seal each individual item in your possession with plastic and duct tape. Memorize the outlines of various planes in the United Terrorist Air Force . . .

Severe: Memorize and be prepared to quote Revelations Chapter 6, Verses 1–8 . . . Contact all prospective sex-partners and make last-chance arrangements. Yes,

> including Janine . . . Prepare bite-sized portions of cyanide in case your underground bunker is penetrated, pussy.[66]

You could argue that these examples are marginal. But commentary at the edge of society is always "marginal." We tend to underestimate it, just like we tend to believe in a clear, simply structured relationship between the edge and the center of society. For the comedic game of infection is hardly limited to Alice Cooper and a few cyberpunks. Just as "marginal," but in fact fully integrated into the cultural industry, are, for instance, the new computer game *.hack/INFECTION* for PlayStation 2, James Cunningham's animation film *Infection* (New Zealand's entry in competition for the 2000 Cannes Film Festival), or Danny Boyle's *28 Days Later,* released in 2003.[67] The film tells the story of the "rage" virus, under experiment in the laboratory to develop drugs against human anger. When infected laboratory monkeys are freed by radical animal rights activists, the rage infection spreads to all of humankind and transforms the world into a hell of aggression—a dystopic vision of plague that in the context of the "bioterror" of the radical activists simply emphasizes how much hidden "rage" there is in the world we live in. It is none other than our own aggressiveness come back to destroy "humankind" in the form of infected monkeys.[68]

This plot is reminiscent of Terry Gilliams's 1995 film *Twelve Monkeys,* which is worth a closer look. *Twelve Monkeys* is a complex vision of a virus that has managed to exterminate nearly all of humankind. It forces the few remaining survivors to live underground, while the animals alone rule above. At the same time, the film is also the dream that the protagonist, James Cole, has as a small boy, who in the dream witnesses his own death thirty years later: he will be shot as he returns from the future to the 1990s to prevent a stranger from setting the killer virus free. When Cole first appears in the past, he is committed as a mental patient to a psychiatric clinic, where he wants to telephone to fulfill his mission. What an idea! The manic Jeffrey, who explains the world of the mad to Cole, warns him: "A telephone call? That's communication with the outside world! Doctor's discretion. Hey, if alla these nuts could just make phone calls, it could spread. Insanity oozing through telephone cables, oozing into the ears of all those poor sane people, infecting them! Whackos everywhere! A plague of madness."[69]

But madness is not the only infection that threatens. Cole talks nonstop about the virus that in the near future is going to wipe out humankind with the exception of a few survivors—the discourse of a lunatic, as he himself soon realizes. And Jeffrey, the sublime flake, son of a famous virologist and a radical animal rights activist, even manages to make Cole believe that "maybe I'm the one who wiped out the human race? It was my idea?" Or was it Jeffrey's "plan"? Years later, when Cole appears for the second time in the past, Jeffrey and his Army of the Twelve Monkeys free the animals in the New York Zoo, in order to lock his father—who performs animal experiments in his virology laboratory—in the cage in their place. Cole no longer knows in which world and in which time he is living. Together with his psychiatrist, Dr. Railly, whom he kidnaps, he flees into a movie theater to see Hitchcock's *Vertigo* and *The Birds* and to discover that film and reality are beginning to blur: "Funny, it's like what's happening to us, like the past." Is the past nothing more than the films a man saw in his boyhood? At the end, which in the film is also the beginning, we see the virologist's laboratory assistant who has just managed to board a plane with the lethal superbug in his hand baggage. He is on his way to the Third World—Rio de Janeiro, Karachi, Beijing—to release the virus.

Twelve Monkeys is a subversive bioterror fantasy that reveals some of the peculiarities of all bioterror fantasies. "Biohazard," as mockingly proclaimed on the taped mouth of the kidnapped virologist, is the game played in an urban world full of filth and chaos, crazy Jeffrey's "brilliant idea." And yet it is nothing more than a game—the empty revolutionary rhetoric of a gang of unwashed animal liberationists who would not have a clue how to get their hands on the viruses of the famous dad even if they wanted to. The assistant, on the other hand, a lone, fair, redheaded, nontalkative type, about whom we learn absolutely nothing, will do "it"—the ultimate bioterrorist raid on humankind—without cause. That is, he does the deed simply because it is possible—but possible only in the laboratory of an American Nobel prize winner, whence he bears the virus into the world. No question: It is "our" virus; so just as the Twelve Monkeys have a harmless obsession with freeing animals, "our" way is to be radical.[70] In other words, the stories about "bioterror" are, as Hegel says, stories about our "own nature." That is the point the film makes, particularly in the time warp that enables an encounter in the airport between James Cole as an eight-year-old boy and himself as a forty-year-old man.

In Chris Marker's short film *La Jetée* (1962), which was the inspiration for Gilliam's film, a man repeatedly dreams of the moment sometime before the Third World War when, as a young boy in the Paris airport terminal, he saw his own execution as an adult. *Twelve Monkeys,* too, depicts a recurring dream of a person's own future death. Why? Is the dream image of one's own death a filmic metaphor for the "legitimate" fear of the nuclear (1962) or bioterrorist (1995) demise of humankind? Or is it more the awful premonition that "destroying humankind" refers to oneself? But "to wipe out the human race" is a totally unacceptable idea, an obscene, taboo desire. According to Freud, dreams represent wish fulfillment and should safeguard sleep. In *Twelve Monkeys,* though, only the subject's awareness of punishment by his own death can save him from waking out of the bioterror dream. By the same token, his desire can only be fulfilled with his death.

Gilliam's *Twelve Monkeys* suggests that in the end, the death impulse means the death "of all." The bioterror dream of the late 1990s, which is still being dreamed today, scarcely goes that far. Who dreams it dreams of order; the wish fulfillment of this dream would not be the eradication of humankind but the five anthrax letters, which have made the dream so true that the dreamer is not forced to wake up and to realize that the whole thing was only his own fantasy.

"Anthrax" Jokes and Hoax Letters

On December 31, 2001, the to-all-appearances right-wing fundamentalist *Online Newspaper Gazette* from Overland Park, Kansas, reported on someone (not mentioned by name) who supplied fake anthrax powder as a "joke item." Ever since the right-wing extremist Clayton Lee Waagner was arrested for having sent hundreds of anthrax hoax letters to abortion clinics in the wake of September 11, the media frenzy around his case resulted in increased demand for false anthrax (hoax or fake anthrax), skyrocketing prices, and delivery difficulties.

Some people looked for alternatives. A local drug dealer told the *Gazette* that some of his customers had even substituted cocaine for "fake anthrax" in the anthrax hoax letters, to get a bit more of an effect than simple powdered sugar: "It is a very expensive alternative but it is cheaper than real anthrax." The journalist also reported from the bizarre world of right-wing extremist militias. For this group, "fake anthrax" was like beginner drugs: "Fake anthrax, whatever the price, is a godsend for mild mannered entry-level terrorists and others wanting to be 'bad boys' without actually endangering anyone. Says one terrorist wannabe 'The older terrorists say fake anthrax is like kissing your sister, but hey, I've got a really hot sister!'"[71]

All wrong. The whole thing is a joke, an avowedly fake site, an "anthrax" farce. Or rather, a metafarce that points up the irony of all the anthrax hoax letters—though their purpose was not at all ironic. Many, such as "domestic" terrorist Waagner, who sent more than five hundred letters to abortion clinics all over the United States, took their threatening hand-waving with false anthrax very seriously; others were only joking, or wanted to settle private scores. We do not really know, since unfortunately these hoax letters are not documented on the Internet. How do the threats work? What are they advertising for? There are thousands of these letters—what are they about? Cultural studies experts are a little thin on the ground both within the FBI and the community of Internet columnists. Who cares what low-brow comedy has to say? Is it not enough that the authors are tracked down and arrested? Do we have to do them the honor of reading their texts?

The only thing that is clear is that in fall 2001 they were all playing the bioterror game. The "anthrax" game was better than any videogame; it was real "biohazard" in the "real world." The sheer numbers of people who mischievously baited the media system with false alarms testifies to an outrageous delight in simulating infection and participating in this massively multiuser online game—relishing infection as radical subversion. Because the phantasm of "bioterror" does not only encode the fear of foreigners and the secret pleasure in killing these vermin and wiping out these microbes, but also the shameless desire to be infected. The dream of order fulfilled through the death of five people found its echo in a dissolute dream of disorder, of brazen laughter, and of games.

Maybe that is what the "anthrax" letters have to tell us. We can only hope that they are preserved for future historians.

Epilogue: Smallpox Liberalism

At the beginning of June 2003 the health authorities in three midwestern states registered nineteen cases of a rarely seen disease: something that looked like smallpox. On June 8 the *Washington Post* reported that the disease was caused by the monkeypox virus *(Orthopoxvirus simiae)*. It was known to crop up occasionally in Africa and Asia but never before in the western hemisphere. The virus is closely related to the agent that causes smallpox, but it is not as infectious or as deadly. The CDC in Atlanta responded immediately with a nationwide warning to all physicians: "We have an outbreak," said James Hughes, the CDC's director. "I'd like to keep it relatively small. I don't want any more cases. We're doing everything we can to try to contain this."[1] The health authorities in Milwaukee and CDC experts attributed the contagion to infected prairie dogs exposed to the bug in exotic pet stores.

In humans, monkeypox produces a zoonotic disease, that is, a condition caused by pathogenic bacteria or viruses that jump the species barrier from animals to human. Other examples include Ebola, West Nile virus, SARS, and probably AIDS. The monkeypox virus outbreak in Wisconsin and Illinois was quickly halted: it did not spread beyond the local area, and only thirty-seven people got sick. Even the spread of information about the first monkeypox outbreak in the history of the United States was limited; in Europe, the short-lived epidemic attracted little, if any, notice.[2]

Monkeypox is less lethal than "proper" smallpox (approximately 10 percent of those infected die, as opposed to 30 percent with smallpox), but it produces similar symptoms. And because it can be transferred easily from one person to the next, it is more dangerous than anthrax. It is reasonable to fear it, yet no panic was forthcoming. The health authorities proved that they can bring a local outbreak of a pathogenic virus under control relatively quickly, and the ruling out of terrorism even seemed to cause some secret disappointment. The infected prairie dogs presented a completely different picture from infected immigrants or bioterrorists. These innocent creatures simply reminded us that we share the planet not only with animals but especially with invisible organisms that occasionally multiply in our bodies and at our expense. We live in an open, unsecured neighborhood with microbes and viruses, and even the animals that we breed and eat sometimes turn into deadly ambassadors from this invisible world. This danger is real: avian flu from China has already arrived in Afghanistan, Croatia, Denmark, France, Greece, India, Israel, Malaysia, Nigeria, Romania, Switzerland, and Turkey, among others, and health experts must now reckon with the threat of a worldwide flu pandemic.

How Real Is the Threat?

The history of the monkeypox outbreak in the summer of 2003 can help us to keep the hype over bioterror in perspective and not lose sight of the relevant political questions. I am neither a microbiologist nor a weapons expert, and I have no access to classified intelligence (although the Internet can be quite revealing). But I have also noticed that those who rightfully claim such expertise differ wildly on the issue of risk. For example, one cautious assessment of thc dangcr of bioterror appeared in *Science* magazine in March 2005 in the form of an open letter to Elias Zerhouni, director of the National Institutes of Health. Signed by some 750 bioscientists, the letter laments the massive siphoning off of government research dollars from civilian infection and immunity to military biodefense. Work on a few hazardous organisms such as plague, smallpox, and anthrax is not only expensive and scientifically unproductive, wrote the scientists. It also spreads dangerous know-how, because "increasing the number of labs and people working on bioterror agents would raise the risk of an accidental release or deliberate attack."[3]

On the other hand, it is obvious that even expert knowledge occasionally falls prey to the hysteria and political phantasms that, inspired by the global war on terror, shape the worldview of many. A good example is Ken Alibek, the renowned former Soviet bioweapons scientist and defector who may be one of the best-informed experts in this field. In a June 2005 interview with the *Neue Zürcher Zeitung*, one of the leading liberal-conservative newspapers in Europe, Alibek referred to "a study by economic consultants in the USA" that "estimates the costs of a smallpox attack on the USA to be 177 billion dollars per week." This figure is so monstrous—beyond any stretch of the imagination—that it bespeaks an equally monstrous threat. But how big *is* the risk? "No one knows," said Alibek, but "we should appreciate that there are certain things Al Qaeda might able to manufacture. Before 2001, nobody could have conceived of the attack on the World Trade Center—and yet it happened. Sometimes we pay a high price for our ignorance. We should be prepared, and not underestimate the danger."

If Alibek's risk analysis is worth as much as his statement that "nobody could have conceived" of the attack on the WTC, then it is time to tone down the excitement. But, asked the *NZZ*, what about unemployed Soviet bioweapons engineers? Couldn't Al Qaeda have hired them? It is an oft-asked question, and Alibek dismissed it:

A: To be honest, I don't give these stories much credit. What if I told you Swiss scientists are paid by Al Qaeda? You could believe it or not. It has become somewhat fashionable to disparage Russian scientists. Americans, Iraqis, or whoever could just as well be involved with Al Qaeda. Why doesn't anyone speculate about that?

Q: But could one of your students build a biological weapon in the garage?

A: Let me reply philosophically: Two hundred years ago, it was unthinkable to believe that people would be using mobile telephones, wasn't it? Everything changes. Our knowledge grows, and technology develops incredibly quickly. These days even high-school kids can breed recombinant microbial strains. I am not saying that a student is in a position to build a biological weapon all by himself. But the knowledge needed to do it is certainly there.

This desultory musing continued in a lively fashion. At one point, contrary to the statements of FBI experts, Alibek stated that the anthrax powder used for the U.S. attacks in October 2001 was possibly manufactured "somewhere in the forest, in a car, without a microscope," which just goes to show where even expert discourse can lead: into a quagmire of speculation and phantasms.[4]

For me, the question is not one of trying to atomize the danger from worldwide research on bioweapons. What is really worrisome is that today, according to a number of studies, the United States primarily but also many other countries are engaged in extensive so-called biodefense research. It is entirely plausible that in ten to twenty years the weapons for the battlegrounds of the future will have dramatically changed and that a large number of lethal and nonlethal biological weapons will be among them. The October 2005 announcement, for instance, that U.S. Army and other scientists had successfully reconstructed not only the genome of the flu virus from 1918, which killed some 25 million people, but the live virus itself, is scary.[5]

In view of these realities, it is worth noting that the phantasm of "bioterror" is mainly about something else. "Bioterror" denotes the threat from terrorist "others," who today more and more are identified with the inscrutable Al Qaeda. Their representation in the media has its apogee in the fantasy image of infected Muslim martyrs: "bioweapons" par excellence whose infiltration of "our" world was foreshadowed by the poison letters. This phantasm was extremely effective in unleashing the global war on terror; yet it obscures the unsettling fact that only in "our" laboratories could real superbugs be incubated and released. That is the lesson of *Twelve Monkeys*, and today there is no reason to doubt it. It may turn out that Al Qaeda did manage to obtain material for a dirty bomb, or tried to, and that it experimented with chemical weapons.[6] But the claim that Al Qaeda has the wherewithal to carry out even a halfway efficient bioweapons attack is both science fiction and disinformation.

The Smallpox Model of Power

No matter how you estimate the risks of bioweapons and bioterror, as well as the more commonplace risks of infection in the context of global streams of migration, it is only one side of the coin. Another question is what political conclusions one draws from such estimates, and which political strategies one develops to contain the risk of infection. Is it reasonable to prepare for the worst in every case, and to anticipate every possibility of infectious disease? Perhaps the authorities of seventeenth-century European cities were right to institute methods of quarantine and monitoring to prevent future outbreaks of plague from killing entire populations. But is there, in an age of globalized infection, an alternative to Foucault's plague model of power that does not end up either in the dream of order or in subversive farce?

In the late 1970s, Michel Foucault re-examined and revised his then still very radical political positions in a way that is interesting even today. In a discussion with the "antipsychiatrist" David Cooper and others, Foucault observed in 1977:

"But, in the end, I've become rather irritated by an attitude, which for a long time was mine, too, and which I no longer subscribe to, which consists in saying: our problem is to denounce and to criticize; let them get on with their legislation and their reforms. That doesn't seem to me the right attitude."[7] Foucault wanted to overhaul the political model that he had formulated in *Discipline and Punish,* to wit, that the only possible response to a disciplining power is farce and the laughter of the infected. Foucault now recognized that government has genuine administrative problems that must be solved, and in that context even the discourse of reform can be meaningful.

This shift in his political posture was also reflected in a new analysis of the forms of government of the modern age, which deviates from the disciplinary model and which for our purposes is very instructive. In his 1978–79 lectures on the *History of Governmentality,* published in 2004, Foucault posited a modern governmental rationality in which individual freedom constitutes an irreducible entity, something "absolutely fundamental."[8] Modern governmentality is a form of power that depends on the freedom of individuals for its functioning. In turn, this freedom amounts to a limit on the power of the state. Although the individual freedom might not be a given but a product—a tactic, you might say—of the liberal state, there is no way around it: in modernity, individual freedom becomes a limit for governmental power.

Thus Foucault, the radical critic of power, recognized in liberalism that form of national rule in which power does not claim every part of an individual down to the least elements of his body and the deepest feelings of his soul. Rather, power must be considered to be limited by the individual freedom on which the system turns. Foucault's thought model for liberal government reform, as he envisioned it, is the state's handling of smallpox in the eighteenth century. In contrast to the disciplinary form of defense in the case of plague—which still has currency as a political model, as we have seen—the French authorities of the eighteenth century (Foucault talks de facto of the end of the Ancien Régime) reacted to smallpox primarily via statistical observation. That is, they measured the actual number of medical cases and tried to protect the population empirically by inoculating them against infection.

In the face of smallpox, the "fundamental problem is to know how many people are afflicted with smallpox, with what damage and sequelae . . . what risks people take when they allow themselves to be vaccinated, how high the probability is that an individual will die or become ill of the pox despite vaccination, what the statistical effects on the population are."[9] However, in liberal government, risk management that takes these problems into account should not go so far that it tips over into disciplining individuals. An overly strong state destroys its own goals—the modern state must, rather, respect the relative "impenetrability" of society.[10] Indeed, liberals have always known that, and Benjamin Franklin is claimed to have remarked: "Those who would give up essential Liberty, to purchase a little temporary Safety, deserve neither Liberty nor Safety." But Foucault's reflections are interesting and important, because they demonstrate quite plainly a fundamental fact: The liberty of individuals must be respected *even at the cost of a certain risk of infection.*

Terrorism Is No Virus

The smallpox model of power, as Foucault conceptualizes it, abandons the dream of totally eradicating pathogens, invaders, and disease germs. Power is described in terms both epidemiologically enlightened and politically liberal as coexisting with the pathogenic invader, understanding its prevalence, gathering data, generating statistics—and of course also launching "medical campaigns" which may include some form of classifying and controlling individuals.[11] But—and this point is crucial—discipline, not to mention total discipline, is no longer a reasonable objective of liberal power in the modern age. In other words, in the light of Foucault's two models, which I have drawn on in this book for my analysis, it becomes clear that current policy in the age of the global war on terror seems to be vacillating between remaining with the liberal smallpox model of partial coexistence with aliens or switching over to the plague model of total control and discipline.

The smallpox model teaches us that we can live not only with infectious germs—a point Joshua Lederberg emphasizes—but probably equally well with terrorism: as a form of political criminality that should be fought with all the weight of the law, through appropriate policing, as a tactical problem of postmodern society, not a strategic one.[12] Paradoxically, it could even be that the attacks on London of July 7, 2005, signaled an end to the global war on terror, or at least ushered in a new, more moderate European perspective on this "war." For the local police work of Scotland Yard was quick and efficient, and its findings were sobering. The attackers grew up in the naturalized Pakistani middle class in Leeds, and neither the authorities nor their neighbors or friends ever pegged them as extremist or inclined to violence. They were proud to be British, played cricket, and one of them was deeply involved as a teacher and social worker for troubled youth. Thus it was evident, at least to Europeans, that these young people belonged to our world.

But to what extent exactly? Were they "sleepers"? That is, coldly calculating enemies, recruited long ago and living for years under cover? Were they like the former agents of the Cold War who trained patiently for a career in the government bureaucracy of the enemy only to betray him once they had arrived decades later? The circumstances of Leeds tell a different story. The perpetrators were normal, native Britishers, fairly well integrated boys, like countless others. They were more interested in sports than in politics, but in the last two years were said to be increasingly drawn to religion. They were neither enemies nor foreigners—no "foreign bodies"—yet they ended up channeling all their energy into causing maximal damage to the society they grew up in. After coming under the influence of fundamentalist religion, they were recruited at some point, by unknown persons, probably not too long before the attacks. From that time on, they were "strangers" in their community, mortal enemies of a way of life that previous to that time had been their own.

What is to be done with them? In the context of postmodern, multicultural immigrationist societies, how do we deal with the danger of young Muslim men being "turned around" in this way? How are our open, pluralistic societies to react to threats of terror? There are basically two possibilities, as we have seen: In the light of smallpox liberalism, we are forced to admit that defense possibilities beyond the maintenance of liberal culture and effective police work are relatively small. We have to live with the risk of "infection" (which is, as the example shows, quite more religious than "terrorist"). A second and alternative, but perhaps obvious, suggestion (mentioned earlier) was put forward in the *Washington Post* on August 23, 2005, by Paul Stares and Mona Yacoubian, both associated with the United States Institute of Peace: Let us understand terrorism as a "virus" that releases an "infection." According to the authors, this approach has three advantages:

> First, it would encourage us to ask the right questions. What is the nature of the infectious agent, in this case, the ideology? Which transmission vectors—for example, mosques, madrassas, prisons, the Internet, satellite TV—spread the ideology most effectively? . . . Second, an epidemiological approach would help us to view Islamist militancy as a dynamic, multifaceted phenomenon. Just

> as diseases do not emerge in a vacuum but evolve as a result of complex interactions between pathogens, people, and their environment, so it is with Islamist militancy . . . Third, it would encourage us to devise a comprehensive, long-term strategic approach to countering the threat. Public health officials long ago recognized that epidemics can be rolled back only with a systematically planned, multi-pronged international effort.[13]

The particular appeal of metaphors, and especially of well-intentioned metaphors, is that they ring "true." Given that terrorism appears to be spreading worldwide like a "disease," then surely it makes sense to look at the predisposing factors for it from the vantage of epidemiology. And does it not follow that, in the age of the war on terror, policy ought to be thought of as epidemic control? What could be more practical than using a "viral" model to describe the spread of Islamic ideology and Al Qaeda's new forms of Internet communication? The problem with these proposals is twofold.

First, to say that ideas spread "virally" is a little silly, because all cultures work that way. By definition, culture involves copying, duplication, and uncontrolled dispersion. Signs and symbols circulate and propagate more or less rapidly according to the constraints of the dominant medium, and usually over very wide areas. Traditional cultures are "infected" with new ideas, and signifiers evolve in such a way that they only make sense within the network, that is, in conjunction with other signifiers. An example of the diffusion of political culture through the global media is the one I have analyzed in this book: the metaphorical and metonymic shifts of the "anthrax" signifier. In the space of half a year, "anthrax" transmogrified from five letters into an Iraqi nuclear, biological, and chemical weapons program threatening the entire world. People can call it "viral," if they wish; but in our case, we have to be careful not to confound the analytical tool and the analyzed object.

Instead of cloaking such shifts in biologistic metaphors, it is probably more helpful to analyze them from the perspective of cultural studies. Viewed that way, not much is gained by using biological language to reiterate for the umpteenth time that the Internet is a web. We all know that the communications capability of the Internet can serve to undermine the "panoptic" control of a strongly hierarchical, centralized society and power structure, and to establish "dangerous" contacts. But that tells us nothing about the best strategy for countering terrorism.

Stares and Yacoubian suggest instead what appears to be quite a sophisticated epidemiological model. They seem to recognize that it is not simply a question of "destroy[ing]" the enemy, as Robert Koch, the father of bacteriology, said in his anthrax essay of 1876.[14] They also seem to know that, as was acknowledged in the 1920s, agents of infectious disease have long been and continue to be a part of our own biological surroundings.[15] We coexist with some of them by protecting ourselves locally, as it were, for example, by immunizing against them. We lessen the ability of others to survive by altering their environment. That is the sort of model Stares and Yacoubian appear to be proposing. Yet it is not altogether clear whether they really believe that political epidemic control has a hope of triumphing over terrorism. They are careful to say only that their approach might help to reverse "the worrisome trends."

But metaphors are slippery aids to comprehension, especially where politics is concerned. Because—and this is my second point—as soon as one takes the "viral" metaphor so seriously that it begins to dictate the logic of the *real* global defense and counterstrategies, it becomes dangerous. Stares and Yacoubian speak of initiating an aggressive strategy of propaganda in the cultural milieus of Arabic countries where terrorism arises. It would consist primarily of "*cleansing* the most hate-filled *vectors*." Now, in the jargon of epidemiology, "vectors" are the carriers or carrying medium of disease, the *Anopheles* mosquito in the case of malaria, say, or rats in the case of plague. In other cases, for example, AIDS, the vectors would be the exchange of blood between infected people, and finally the infected persons themselves. "Cleansing" these "vectors" means either fighting them virtually to extinction—or, if that isn't possible, pursuing rigorous containment with quarantine measures and the like. "Hate-filled vectors" is therefore a metaphorical formulation that directly evokes the worst crimes of the twentieth century—for example, the identification of Jews (with their rumored "hatred" of the German "people") with lice, which were the animal vector for the typhoid fever endemic in certain parts of Eastern Europe. That between 1942 and 1945 the German "national body" "disinfected" itself of these "lice" in the gas chambers was very closely bound up with this metaphor.

Language is not harmless, and we should remember that it has a history. In that sense, Stares and Yacoubian's recommendation to use the virus metaphor as a jumping off point for political strategies in the battle against terrorism is way off the mark. They favor "devising an ideological antidote to neutralize the most infectious tenets of Islamist militancy" and "exploiting the ideological contradictions or schisms within the militant Islamist movement to foment internal dissension and defection." Moreover, they advocate "seek[ing] to immunize vulnerable populations by promoting a moderate counter-ideology that offers a positive, more compelling view of the future." That sounds friendly enough—"moderate"—but it would be nice to know what differentiates this model of ideological warfare from similar—and, incidentally, unsuccessful—attempts during the Cold War.

Perhaps only the lack of respect for the opponent. Radio Free Europe had to be broadcast from West Germany because influencing the media in countries under Soviet rule was unthinkable. Yet Stares and Yacoubian fancy that today, in so-called "high-risk countries and communities of the Muslim world," such ideological interventions might actually take place at the heart of these societies, namely, in "such critical areas as religious and educational institutions, community centers and mass media outlets."[16] The concept of this U.S. cultural imperialism is by no means as sophisticated as the authors would like to suggest with their epidemiologically inspired language. Nor is it pacific. Such an "epidemiological" defense strategy would in fact be a major covert operation, a sort of "cultural" covert action with the goal of massively influencing foreign societies. And it is the virus metaphor itself that suggests these imperialistic interventions. For how can an epidemic be successfully fought if not by "undertak[ing] remedial initiatives that address the key environmental conditions underlying the spread of" typhoid fever, plague, avian flu, or "islamist militancy"?[17] One must clean where one finds dirt.

The recommendations of Stares and Yacoubian show clearly how quickly metaphors can progress from innocuous descriptive language to deadly weapons. Hence, an American attempt to "install democracy" not only in Iraq but in terms of mounting a "global counterterrorism campaign inspired by classic counter-epidemic measures" in all Islamic countries where terrorism "grows," as George W. Bush puts it, would be nothing less than the agenda for a truly global war.

Notes

Prologue: Ground Zero

1. *Superhuman: Killers into Cures*, vol. 2, VHS, produced by Michael Mosley (London: BBC, 2000).

2. Judith Miller, Stephen Engelberg, and William Broad, *Germs: Biological Weapons and America's Secret War* (New York: Simon and Schuster, 2001).

3. Richard Preston, *The Cobra Event* (New York: Ballantine, 1997), p. 38f.

4. Sheryl Gay Stolberg, "A Nation Challenged: The Biological Threat," *New York Times*, September 30, 2001.

5. Associated Press, "White House Faces Disclosure Suit: Group Says Government Had Braced for Anthrax Attacks," *Washington Post*, June 8, 2002, p. A11.

6. James M. Hughes and Julie Louise Gerberding, "Anthrax Bioterrorism: Lessons Learned and Future Directions," *Emerging Infectious Diseases* 8, no. 10 (2002).

7. James J. Bono, "Science, Discourse, and Literature," in *Literature and Science: Theory and Practice*, ed. Stuart Peterfreund

(Boston: Northeastern University Press 1990), pp. 59–89, quote p. 61.

8. Jacques Lacan, *The Psychoses, 1955–1956,* trans. Russell Grigg (New York: W. W. Norton, 1997), p. 224. Thanks to Johannes Fehr, Zurich, for this reference.

9. Slajov Žižek, *Welcome to the Desert of the Real: Five Essays on September 11 and Related Dates* (London: Verso, 2002), p. 47.

10. For the term *phantasm* as used here, see: Jacques Lacan, *The Four Fundamental Concepts of Psychoanalysis: Seminar XI*, trans. Alan Sheridan, ed. Jacques-Alain Miller (New York: Norton, 1978), p. 66.

11. Fritz Göttler, "Man nennt es Morganröte," *Süddeutsche Zeitung*, December 16, 2003.

1. Video Games, 9/11, and the Anthrax Letters

1. Elie Metchnikoff, "Sur la lutte des cellules de l'organisme contre l'invasion des microbes (théorie des phagocytes)," *Annales de l'Institut Pasteur* 1 (1887): 331–336.

2. William Hunter, "The History of Video Games: From 'Pong' to 'Pac-Man,'" *Designboom.com,* September 2000, http://www.designboom.com/eng/education/pong.html.

3. Julius Cohnheim, *Neue Untersuchungen über die Entzündung* (Berlin: August Hirschwald, 1873).

4. See http://www.gamefaqs.com/console/psx/review/R5340.html or http://faqs.ign.com/articles/381/381940p1.html for descriptions of the *Parasite Eve* role-playing computer game.

5. Evelyn Fox Keller, "Der Organismus: Verschwinden, Wiederentdeckung und Transformation einer biologischen Kategorie," in *Vermittelte Weiblichkeit: Feministische Wissenschafts- und Gesellschaftstheorie,* ed. Elvira Scheich (Hamburg: Hamburger Edition, 1996), pp. 313–334.

6. Paula A. Treichler, *How to Have Theory in an Epidemic: Cultural Chronicles of AIDS* (Durham: Duke University Press, 1999), p. 207.

7. See http://gamefaqs.com/console/psx/review/R5340.html.

8. See http://www.darwinia.co.uk/index.html.

9. Timothy Lenoir, "All But War Is Simulation: The Military-Entertainment Complex," *Configurations* 8 (2000), pp. 289–336.

10. See http://www.americasarmy.com.

11. David Ignatius, "Outgaming Osama," *Washington Post,* December 6, 2002, p. A45.

12. Benjamin Pimentel, "Software Simulates Terror Hit," *San Francisco Chronicle,* August 19, 2002.

13. See http://www.capcom.com. "Command+Conquer, Generals," on http://www.gamespot.com/gamespot/gameguides/pc/generals/8.html.

14. "Welcome to the desert of the real" is from *The Matrix,* DVD, directed by Andy Wachowski and Larry Wachowski (1999; Burbank, CA: Warner Home Video, 1999).

15. Anselm Kiefer, cited in Elisabeth Bronfen, "Der unsagbare Kern," *Die Tageszeitung,* October 16, 2001, p. 15.

16. Jacques Derrida, *From Rogues: Two Essays on Reason,* trans. Pascale-Anne Brault and Michael Naas (Stanford, CA: Stanford University Press, 2005), p. 103.

17. Slajov Žižek, *Welcome to the Desert of the Real: Five Essays on September 11 and Related Dates* (London: Verso, 2002), p. 47.

18. Karl Marx, "The Eighteenth Brumaire of Louis Napoleon," part 1, in *On Revolution,* (vol. 1 of *The Karl Marx Library*), ed. and trans. Saul K. Padover (New York: McGraw-Hill, 1971), p. 255.

19. Sandra Silberstein, *War of Words: Language, Politics, and 9/11* (London: Routledge, 2002), p. xii.

20. Ibid., pp. 79–83.

21. "Person-on-the-street" interview captured on September 11 in New York, soon after the attack, cited in ibid., pp. 66, 76.

22. Žižek, *Desert of the Real,* p. 14.

23. *Fight Club,* DVD, directed by David Fincher (1999; Los Angeles: Twentieth-Century Fox, 2000).

24. Hans Ulrich Gumbrecht, "In eine Zukunft gestossen: Nach dem 11. September 2001," in *Terror im System: Der 11. September und die Folgen,* ed. Dirk Baecker, Peter Krieg, and Fritz B. Simon (Heidelberg: Carl-Auer-Systeme, 2002), pp. 100–109; reprinted from *Merkur* 55 (2001): 1048–1054, citation p. 104.

25. Sigmund Freud, *Collected Papers,* vol. 4 (New York: Basic Books, 1959), p. 398.

26. Jean Baudrillard, *The Spirit of Terrorism and Requiem for the Twin Towers* (London: Verso, 2002), p. 5.

27. See Jean Baudrillard, "L'esprit du terrorisme," *Le Monde,* November 2, 2001.

28. Navid Kermani, "Dynamite of the Spirit," *Times Literary Supplement,* March 29, 2002, p. 15.

29. Ibid.

30. Olivier Roy, *Globalized Islam: The Search for a New Ummah* (New York: Columbia University Press, 2005), cited in Max

Rodenbeck, "The Truth about Jihad," *New York Review of Books,* vol. 52, no. 13, August 11, 2005, p. 53.

31. Kermani, "A Dynamite of the Spirit," p. 15.

32. Ibid.

33. Žižek, *Desert of the Real,* p. 16.

34. This version of the events is abbreviated inasmuch as I have left out the seventeen nonlethal cases of anthrax; for a detailed account see Leonard A. Cole, *The Anthrax Letters: A Medical Detective Story* (Washington, DC: Joseph Henry Press, 2003).

35. See the report of Otto Lummitsch, one of the witnesses to the preparation of the gas attack at Ypres in February 1915: "In February the Supreme Army Command directed the decampment to the front of the first gas divisions in the so-called Disinfection Campaign"; cited in Dietrich Stoltzenberg, *Fritz Haber: Chemist, Nobel Laureate, German, Jew: A Biography* (Philadelphia: Chemical Heritage Foundation, 2005).

36. Peter Sloterdijk, *Luftbeben: An der Quelle des Terrors* (Frankfurt/M: Suhrkamp, 2002), pp. 27ff. Rudolf Walther, "Terror, Terrorismus," in *Geschichtliche Grundbegriffe: Historisches Lexikon zur politisch-sozialen Sprache in Deutschland,* vol. 6, ed. Reinhart Koselleck et al. (Stuttgart: Klett-Cotta, 1990), pp. 323–444.

37. Judith Miller, Stephen Engelberg, and William Broad, *Germs: Biological Weapons and America's Secret War* (New York: Simon and Schuster, 2001), pp. 38–39.

38. For a comprehensive overview, see Joseph Cirincione, with John B. Wolfsthal and Miriam Rajkumar, *Deadly Arsenals: Nuclear, Biological, and Chemical Threats,* 2nd ed. (Washington, DC: Carnegie Endowment for International Peace, 2005).

39. Robert Koch, "Etiology of Anthrax," in *Essays of Robert Koch,* trans. K. Codell Carter (New York: Greenwood Press, 1987).

40. Gerald L. Geison, *The Private Science of Louis Pasteur* (Princeton: Princeton University Press, 1995), pp. 145–176.

41. H. Caksen, F. Arabaci, M. Abuhandan, O. Tuncer, and Y. Cesur, "Cutaneous Anthrax in Eastern Turkey," *Cutis* 67 (2001): 488–492.

42. See the World Health Organization's *World Anthrax Data Site:* http://www.vetmed.lsu.edu/whocc/mp_world.htm.

43. Erin Harty, "Anthrax Invades Texas," *VetCentric,* July 25, 2001, http://www.vetcentric.com/magazine/magazineArticle.cfm?ARTICLEID=1399.

44. See the Centers for Disease Control and Prevention (CDC) website on anthrax: http://www.bt.cdc.gov/agent/anthrax/index.asp.

45. Federal Bureau of Investigation, "Photos of Anthrax Letters to NBC, Senator Daschle, and NY Post," national press release, October 23, 2001.

46. See Douglas Rushkoff, *Media Virus! Hidden Agendas in Popular Culture,* 2nd ed. (New York: Ballantine Books, 1996).

47. http://www.cnn.com/SPECIALS/2001/trade.center/anthrax.section.html>Map: Anthrax cases around the world.

48. http://www.pbs.org/wgbh/nova/bioterror/agents.html.

49. Barbara Hatch Rosenberg, Federation of American Scientists, "Analysis of the Anthrax Attacks," news release, February 5, 2002.

50. For a list with links to reports of hoax letters see http://www.anthraxinvestigation.com/writing1.html#hoax.

51. See http://www.anthraxinvesgitation.com/HoaxVsReal.html.

52. See, for example, Law Enforcement Agency Resource Network, Beyond Anthrax: Extremists and the Bioterrorism Threat, http://www.adl.org/learn/Anthrax/Hoaxes2.asp?xpicked=3&item=6 and http://www.adl.org/learn/Anthrax/Harris.asp? xpicked=3&item=5.

53. Laura Snyder and Jason Pate: "Tracking Anthrax Hoaxes and Attacks," Center for Nonproliferation Studies, http://cns.miis.edu/pubs/week/020520.htm.

54. President George W. Bush, address, "Remarks Following Discussions with Prime Minister Tony Blair of the United Kingdom and an Exchange with Reporters, November 8, 2001," *Weekly Compilation of Presidential Documents* 37, no. 45 (November 12, 2001): 1610.

2. Bioterror and Weapons of Mass Destruction

1. Jacques Lacan, *The Psychoses, 1955–1956,* trans. Russell Grigg (New York: W. W. Norton, 1997), p. 224.

2. Jacques Lacan, "The Instance of the Letter in the Unconscious, or Reason since Freud," in *Ecrits: A Selection,* trans. Bruce Fink (New York: W. W. Norton, 2002), p. 141.

3. Jacques Derrida, "Difference," in *Margins of Philosophy,* ed. Jacques Derrida (Chicago: University of Chicago Press, 1982), p. 3–27.

4. *Anthrax,* "Anthrax (the band) vs. Anthrax (the disease)," press release, October 10, 2001.

5. President George W. Bush, "Address before a Joint Session of

the Congress on the United States Response to the Terrorist Attacks of September 11, September 20, 2001," *Weekly Compilation of Presidential Documents* 37, no. 38 (September 24, 2001): 1347–1351.

6. *Neue Zürcher Zeitung,* Research and Technology [Forschung und Technik] section, January 29, 2003.

7. President William J. Clinton, "Interview with Judith Miller and William J. Broad of the New York Times, the Oval Office, January 21, 1999," *Weekly Compilation of Presidential Documents* 35, no. 4 (February 1, 1999): 109–115.

8. Slajov Žižek, *Welcome to the Desert of the Real: Five Essays on September 11 and Related Dates* (London: Verso, 2002), p. 36.

9. See United Kingdom, Office of the Prime Minister, "Part 2: History of UN Weapons Inspections," *Iraq's Weapons of Mass Destruction: The Assessment of the British Government* (London: HMSO, 2002).

10. Raphael Perl, *Terrorism and National Security: Issues and Trends,* CRS Report for Congress, June 8, 2005.

11. See *National Defense Authorization Act for Fiscal Year 1997,* Public Law 104-201, Title XIV, "Defense against Weapons of Mass Destruction," 104th Congress, 2nd session (September 23, 1996).

12. Jan van Aken, "Saddams Phantompocken," *Frankfurter Allgemeine Zeitung,* February 21, 2003, p. 44.

13. Kurt Langbein, Christian Skalnik, and Inge Smolek, *Bioterror: Die gefährlichsten Waffen der Welt* (Stuttgart: Deutsche Verlags-Anstalt, 2002), p. 42.

14. Malcolm Dando, *Bioterrorism: What Is the Real Threat?* (Bradford Science and Technology Report No. 3, 2005), p. 10.

15. See Bush, "Address" (September 24, 2001): 1347–1351.

16. Massimo Calabresi and Sally Donnelly, "Cropduster Manual Discovered," *Time Online Edition,* September 22, 2001.

17. President George W. Bush, address, "The President's News Conference, October 11, 2001," *Weekly Compilation of Presidential Documents* 37, no. 41 (October 15, 2001): 1452–1462.

18. President George W. Bush, address, "USA Patriot Act, Remarks, October 26, 2001," *Weekly Compilation of Presidential Documents* 37, no. 43 (October 29, 2001): 1551.

19. See President George W. Bush, "The President's Radio Address, November 3, 2001," *Weekly Compilation of Presidential Documents* 37, no. 45 (November 12, 2001): 1600–1601.

20. President George W. Bush, address, "Satellite Remarks to the Central European Counterterrorism Conference, November 6, 2001," *Weekly Compilation of Presidential Documents* 37, no. 45 (November 12, 2001): 1604.

21. President George W. Bush, address, "Remarks Following Discussions with Prime Minister Tony Blair of the United Kingdom and an Exchange with Reporters, November 8, 2001," *Weekly Compilation of Presidential Documents* 37, no. 45 (November 12, 2001): 1611.

22. National Security Advisor Condoleezza Rice, "Press Briefing on the President's upcoming visit to the United Nations," press release, February 8, 2001.

23. President George W. Bush, address, "Remarks to the United Nations General Assembly in New York City, November 10, 2001," *Weekly Compilation of Presidential Documents* 37, no. 46 (November 19, 2001): 1638–1639.

24. President George W. Bush, address, "Remarks at a Welcoming Ceremony for Humanitarian Aid Workers Rescued from Af-

ghanistan and an Exchange with Reporters, November 26, 2001," *Weekly Compilation of Presidential Documents* 37, no. 48 (December 3, 2001): 1713.

25. President George W. Bush, address, "Address before a Joint Session of the Congress on the State of the Union, January 29, 2002," *Weekly Compilation of Presidential Documents* 38, no. 5 (February 4, 2002): 135.

26. Bob Woodward, *Plan of Attack* (London: Pocket Books, 2004), p. 30. Glenn Kessler, "U.S. Decision on Iraq Has Puzzling Past. Opponents of War Wonder When, How Policy Was Set," *Washington Post,* January 12, 2003.

27. Neil Mackay, "Former Bush Aide: US Plotted Iraq Invasion Long before 9/11," *Sunday Herald,* Scotland, January 11, 2004.

28. Richard A. Clarke, *Against All Enemies: Inside America's War on Terror* (New York: Free Press, 2004), p. 32.

29. See the very informative article by Robert Dreyfuss on the leader of the Iraqi National Congress and favored associate of the neoconservative Achmed Chalibi: "Thinker, Banker, NeoCon, Spy," *American Prospect,* vol. 13, no. 21, November 18, 2002.

30. Clarke, *Against All Enemies,* pp. 227–232.

31. http://www.cbsnews.com/stories/2002/09/04/september11/main520830.shtml.

32. Dan Barry, "A New Account of Sept. 11 Loss, with 40 Fewer Souls to Mourn," *New York Times,* October 29, 2003.

33. Marc Danner, "The Secret of War: The Downing Street Memo," *New York Review of Books,* vol. 52, no. 10, June 9, 2005, pp. 70–73.

34. See President George W. Bush, "Address before a Joint Ses-

sion of the Congress on the State of the Union, January 29, 2002," *Weekly Compilation of Presidential Documents* 38, no. 5 (February 4, 2002): 133–139.

35. *National Defense Authorization Act for Fiscal Year 1997,* Public Law 104-201, Title XIV, "Defense against Weapons of Mass Destruction," Section 1402, 104th Congress, 2nd session (September 23, 1996).

36. President George W. Bush, address, "Address to the Nation on Iraq from Cincinnati, Ohio, October 7, 2002," *Weekly Compilation of Presidential Documents* 38, no. 41 (October 14, 2002): 1716–1717.

37. *Neue Zürcher Zeitung,* March 10, 2003, p. 2; a full account in *Neue Zürcher Zeitung,* March 13, 2003, p. 3.

38. Don Kirk, "South Korean Leader Says Move Was Meant to Aid 'Sunshine' Policy: Payment to North Puts Seoul on Defense," *International Herald Tribune,* January 31, 2003, p. 10.

39. CNN International, January 31, 2003, ca. 16.15–16.40 EST.

40. Kane Pryor, "A National State of Confusion: The Bush Propaganda Machine Has Convinced Americans that Saddam and the No-Longer-Mentioned Osama Are the Same Person—and the Polls Prove It," *Salon.com,* February 6, 2003, http://www.salon.com/opinion/feature/2003/02/06/iraq_poll/index_np.html?x.

41. Colin Powell, "U.S. Secretary of State Colin Powell Addresses the U.N. Security Council," press release, The White House, February 5, 2003.

42. Ibid.

43. Mark Huband and Stephen Fidler, "No Smoking Gun," *Financial Times,* June 3, 2003.

44. Thomas Powers, "The Vanishing Case for War," *New York Review of Books,* December 4, 2003.

45. Greg Miller, "Intelligence on Iraq Wrong, General Asserts," *New York Times,* May 31, 2003.

46. Joseph Cirincione, Jessica Tuchman Matthews, and George Perkovich, with Alexis Orton, *WMD in Iraq: Evidence and Implications* (Washington, DC: Carnegie Endowment for International Peace, 2004).

47. Commission on the Intelligence Capabilities of the United States Regarding Weapons of Mass Destruction, *Report to the President, March 31, 2005* (Washington, DC: GPO, 2005), p. 45.

48. See Central Intelligence Agency, *Comprehensive Report of the Special Advisor to the DCI on Iraq's WMD, 30 September 2004* (Washington, DC: GPO, 2004).

49. See Lucien Israël, *Die unerhörte botschaft der Hysterie* (München: Fink, 1987).

50. Scott Ritter and Rivers Pitt, *War on Iraq: What Team Bush Doesn't Want You to Know* (London: Context Books, 2002), p. 39.

51. Emmanuel Todd, *After the Empire: The Breakdown of the American Order* (New York: Columbia University Press, 2004), chapter 6.

52. Thom Shanker and Eric Schmitt, "A Nation at War: Strategic Shift: Pentagon Expects Long-Term Access to Key Iraq Bases," *New York Times,* April 20, 2003.

53. Under Reagan, Abrams backed the invasion of Nicaragua and was accused of turning a blind eye to human rights violations; he later pled guilty to misdemeanors in the Iran-Contra affair. He

headed the Bush administration's National Security Council office for democracy, human rights, and international operations.

3. The Cobra Event

1. Richard Preston, *The Cobra Event* (New York: Ballantine Books, 1998), p. 8.

2. Judith Miller, Stephen Engelberg, and William Broad, *Germs: Biological Weapons and America's Secret War* (New York: Simon and Schuster, 2001), p. 225.

3. "Richard Preston Gives Us the Horror," *Thresher* (1998), *http://www.thethresher.com/richardpreston.html* (this online magazine is no longer available).

4. Miller, Engelberg, and Broad, *Germs.*

5. For this and the accompanying citations, Preston, *Cobra,* pp. 40–53.

6. Ibid., pp. 224–225.

7. Ibid., p. 300.

8. For more information on the Lesch-Nyhan Syndrome, see the National Institute of Neurological Disorders and Stroke Information page at: http://www.ninds.nih.gov/health_and_medical/disorders/lesch_doc.htm.

9. "Richard Preston Gives Us the Horror," *Thresher* (1998).

10. Preston, *Cobra,* p. 235.

11. Ibid., p. 413.

12. Hansruedi Bueler, email message to author, July 14, 2003.

13. Walter Schaffner, email message to author, July 13, 2003.

14. Richard Preston, "Biology Gone Bad," *New York Times,* November 7, 1997.

15. Richard Preston, Statement for the Record before the Senate Judiciary Subcommittee on Technology, Terrorism, and Government Information, and the Senate Select Committee on Intelligence, April 22, 1998.

16. Jonathan Broder, "Clinton, Saddam, and the Hot Zone," *Salon,* November 11, 1997.

17. Martin Schütz, email message to author, July 17, 2003.

18. Preston, "Biology Gone Bad."

19. Preston, *Cobra,* p. 64.

20. Martin Schütz, email message to author, July 17, 2003.

21. WHO provides figures only for the 1980s; see http://www.vetmed.lsu.edu/whocc/iraq.htm.

22. President William J. Clinton, "Interview with Judith Miller and William J. Broad of the *New York Times,* the Oval Office, January 21, 1999," *Weekly Compilation of Presidential Documents* 35, no. 4 (February 1, 1999): 109–115.

23. Richard A. Clarke, *Against All Enemies: Inside America's War on Terror* (New York: Free Press, 2004), p. 32.

24. Tom Clancy, *Rainbow Six* (New York: Berkley, 1998), p. 581.

25. President William J. Clinton, "Interview with Judith Miller and William J. Broad of the *New York Times,* the Oval Office, January 21, 1999."

26. "Richard Preston Gives Us the Horror," *Thresher* (1998).

27. Miller, Engelberg, and Broad, *Germs,* pp. 194–195.

28. The quotes that follow are all from President William J.

Clinton, "Interview with Judith Miller and William J. Broad of the *New York Times,* the Oval Office, January 21, 1999."

29. Miller, Engelberg, and Broad, *Germs,* p. 224.

30. See Halliburton's website: http://www.halliburton.com for specific information on its products and services.

31. President William J. Clinton, "Interview with Judith Miller and William J. Broad of the *New York Times,* the Oval Office, January 21, 1999."

32. Miller, Engelberg, and Broad, *Germs,* p. 238.

33. Noted in the Kean Report: Thomas Kean, *The 9/11 Commission Report: Final Report of the National Commission on Terrorist Attacks upon the United States* (Norton: New York, 2004), p. 140.

34. Richard Dreyfus, "The Phantom Menace," *Mother Jones,* September/October 2000, http://www.motherjones.com/news/feature/2000/09/phantom.html.

35. Clarke, *Against All Enemies,* p. 162ff. William J. Broad and Judith Miller, "Germ Defense Plan in Peril as Its Flaws Are Revealed," *New York Times,* August 7, 1998.

36. Miller, Engelberg, and Broad, *Germs,* pp. 235–241.

37. Broad and Miller, "Germ Defense Plan."

38. Miller, Engelberg, and Broad, *Germs,* pp. 236–237.

39. Ibid., p. 244.

40. Joby Warrick and Joe Stephens, "Before Attack, U.S. Expected Different Hit," *Washington Post,* October 2, 2001.

41. U. S. National Response Team, *Exercise TOPOFF and National Capital Region After-Action Final Report* (Washington, DC: GPO, August 2001), p. 1.

42. Miller, Engelberg, and Broad, *Germs,* pp. 277–279.

43. Amy E. Smithson, "Frequently Asked Questions: Likelihood of Terrorists Acquiring and Using Chemical or Biological Weapons," The Henry L. Stimson Center, http://www.stimson.org/?SN=CB2001121259.

44. Dreyfus, "Phantom Menace."

45. Ibid.

46. President George W. Bush, "Address before a Joint Session of the Congress on Administration Goals, February 27, 2001," *Weekly Compilation of Presidential Documents* 37, no. 9 (March 5, 2001): 351–357.

47. President George W. Bush, address, "Remarks at the National Defense University, May 1, 2001," *Weekly Compilation of Presidential Documents* 37, no. 18 (May 7, 2001): 685–688.

48. U. S. Department of Defense, *Annual Report to the President and the Congress (1999), Chapter 1: The Defense Strategy* (Washington, DC: GPO, 1999).

49. President George W. Bush, "Remarks at the National Defense University, May 1, 2001."

50. Tara O'Toole, Michael Mair, and Thomas V. Inglesby, "Shining Light on 'Dark Winter,'" *Clinical Infectious Diseases* 34 (2002): 972–983. William J. Broad and Judith Miller, "Government Report Says 3 Nations Hide Stocks of Smallpox," *New York Times,* June 13, 1999.

51. "Dark Winter: Bioterrorism Exercise," Andrews Air Force Base, June 22–23, 2001, Final Script, Johns Hopkins Center for Civilian Biodefense Strategies, Center for Strategic and International

Studies, ANSER, & Memorial Institute for the Prevention of Terrorism, 2001, p. 44.

52. Richard E. Hoffman, "Preparing for a Bioterrorist Attack: Legal and Administrative Strategies," *Emerging Infectious Diseases* [serial online], February (2003).

53. *Neue Zürcher Zeitung,* March 13, 2003, p. 9.

54. President George W. Bush, address, "Remarks on Project BioShield in Bethesda, Maryland, February 3, 2003," *Weekly Compilation of Presidential Documents* 39, no. 5 (February 10, 2003): 153–156.

4. What Is an Author?

1. Michel Foucault, "What Is an Author?" in *Textual Strategies: Perspectives in Post-Structuralist Criticism,* ed. Josué V. Harari (Ithaca: Cornell University Press, 1979), p. 148.

2. Ibid.

3. Ibid., p. 151.

4. Ibid., p. 148.

5. Judith Miller, Stephen Engelberg, and William J. Broad, "U.S. Germ Warfare Research Pushes Treaty Limits," *New York Times,* September 4, 2001.

6. U.S. Senate Committee on the Judiciary, Subcommittee on Technology, Terrorism & Government Information and the Senate Select Committee on Intelligence, *Chemical and Biological Weapons Threats to America: Are We Prepared?* 105th Congress, 2nd session, April 22, 1998.

7. Judith Miller, Stephen Engelberg, and William Broad, *Germs: Biological Weapons and America's Secret War* (New York: Simon and Schuster, 2001), p. 296.

8. Miller, Engelberg, and Broad, "U.S. Germ Warfare Research"; Miller, Engelberg, and Broad, *Germs,* p. 298.

9. Barbara Hatch Rosenberg, Federation of American Scientists, "Analysis of the Anthrax Attacks," news release, February 5, 2002.

10. For details, see Alex R. Hoffmaster et al., "Molecular Subtyping of *Bacillus anthracis* and the 2001 Bioterrorism-Associated Anthrax Outbreak, United States," *Emerging Infectious Diseases* 8 (2002).

11. Patrick Martin, "US Anthrax Attacks Linked to Army Biological Weapon Plant," *World Socialist Web Site,* December 28, 2001, http://www.wsws.org/articles/2001/dec2001/anth-d28.shtml.

12. Miller, Engelberg, and Broad, *Germs,* p. 309.

13. Georg Schöfbänker, "Auf der Spur der Anthrax-Briefe," *Telepolis,* December 18, 2001, http://www.heise.de/tp/r4/artikel/11/11371/1.html.

14. William J. Broad, David Johnston, and Judith Miller, "Subject of Anthrax Inquiry Tied to Anti-Germ Training," *New York Times,* July 2, 2003.

15. Ibid.

16. See also Ellen Ray and William H. Schaap, eds., *Bioterror: Manufacturing Wars the American Way* (Melbourne: Ocean Press, 2003).

17. U.S. Mission Geneva, "Press Conference by Ambassador Donald A. Mahley, Special Negotiator for Chemical and Biological

Arms Control Issues," news release, Palais des Nations Geneva, Switzerland, July 25, 2001.

18. Ibid.

19. Broad, Johnston, and Miller, "Subject of Anthrax Inquiry Tied to Anti-Germ Training." Regarding the difficulties of research in the area of bioweapons and the corresponding "emerging technologies," see *Emerging Technologies: Genetic Engineering and Biological Weapons,* the Sunshine Project, Background Paper #12, November 2003.

20. Gary Matsumoto, "Anthrax Powder: State of the Art?" *Science* 302, no. 5650 (November 28, 2003): 1492–1497.

21. Michael D. Lemonick, "Homegrown Terror," *Time,* February 16, 2004, p. 40.

22. David Tell, "Despite the Evidence, the FBI Won't Let Go of Its 'Lone American Theory,'" *Weekly Standard,* April 29, 2002.

23. Dick Cheney, interview by Tim Russert, *Meet the Press,* NBC, September 8, 2002.

24. The most compelling analysis was made by the British Forensic Linguistics Institute: "Report on the Anthrax Envelopes Sent to Senator Tom Daschle and Mister Tom Brokaw." Regarding the whodunit question, see Ed Lake's interesting and detailed investigation: "The Anthrax Cases," http://www.anthraxinvestigation.com.

25. See, for instance, Jerry White, "US Anthrax Attackers Aimed to Assassinate Democratic Leaders. Media silent on Military Links," *World Socialist Web Site,* http://www.wsws.org/articles/2002/jan2002/anth-j23.shtml.

26. Rosenberg, "Analysis of the Anthrax Attacks."

27. Ibid.

28. See http://archives.cnn.com/2002/US/05/22/9.11.warnings.facts/index.html#summary, click on "Timeline" in this article.

29. Ibid.

30. "Report cites warnings before 9/11," *CNN.com,* September 19, 2002, http://www.cnn.com/2002/ALLPOLITICS/09/18/intelligence.hearings/. See also Joint Senate/House Intelligence Committee, *Joint Investigation into September 11th: First Public Hearing,* 107th Congress, 2nd session, September 18, 2002.

31. Steve Fainaru and James V. Grimaldi, "FBI Knew Terrorists Were Using Flight Schools," *Washington Post,* September 23, 2001.

32. "Hijackers Found Welcome Mat on West Coast," *Washington Post,* December 29, 2001.

33. George Wehrfritz, Catharine Skipp, and John Barry, "Alleged Hijackers May Have Trained at U.S. Bases," *Newsweek Web Exclusive,* September 15, 2001, http://www.msnbc.msn.com/id/3668484/site/newsweek.

34. Patrick Martin, "Was the US Government Alerted to September 11 Attack? Part 2: Watching the Hijackers," *World Socialist Web Site,* January, 18, 2002, http://www.wsws.org/articles/w002/jan2002/sept-j18_prn.shtml.

35. Philip Shenon, "Second Officer Says 9/11 Leader Was Named before Attacks," *New York Times,* August 23, 2005; see also Daniele Ganser, "Able Danger Adds Twist to 9/11," *GlobalResearch.ca,* August 29, 2005.

36. Both the Internet and, now increasingly, books abound in conspiracy theories that not only raise questions and speculations—which is not in itself dubious—but that also rely on dodgy

sources and ultimately boil down to the old, antisemitic "Jewish conspiracy." But there is also a whole series of sites that I consider legitimate because their arguments are sound and because they restrict themselves to verifiable sources. One of these is Emperor's New Clothes, which offers a carefully detailed analysis of the gaps in the official version: http://emperors-clothes.com/indict/indict-1.htm.

37. Thomas Kean, *The 9/11 Commission Report: Final Report of the National Commission on Terrorist Attacks upon the United States* (Norton: New York, 2004), p. 457, n. 96.

38. For the relevant passages relating to communications between NORAD and the FAA, see ibid., first chapter.

39. David Ray Griffin, *The 9/11 Commission Report: Omissions and Distortions* (Northampton, MA: Olive Branch Press, 2005), p. 140. Griffin's source is congressional testimony by NORAD's commander Ralph Eberhard in October 2002. Griffin adds that Eberhard was referring to the period *after* the 9/11 attack, and that, at the time, the Kean report was referring to the period *before* the 9/11 attack. However, the commission did not probe the discrepancy, which is why Griffin quotes from a 1998 document warning pilots that "within 10 or so minutes" of erratic behavior, they would have "jet fighters . . . on their tail" (ibid.).

40. Tom Clancy, *Rainbow Six* (New York: Berkley, 1998), p. 846.

41. Michael Rupper, *Crossing the Rubicon: The Decline of the American Empire at the End of the Age of Oil* (Gabriola Island: New Society Publishers, 2004), p. 395.

42. Kean, *9/11 Commission Report,* p. 458, n. 116.

43. For an explanation of this theory, see: http://www . prisonplanet.com/articles/september2004/080904wargamescover htm.

44. Joby Warrick and Joe Stephens, "Before Attack, U.S. Expected Different Hit," *Washington Post,* October 2, 2001. See also http://www.mdw.army.mil/news/Contingency_Planning.html for images of the table top models of buildings

45. Warrick and Stephens, "Before Attack."

46. Peter Eisler, Fred Bayles, and Dan Vergano, "U.S. Cities Brace for the Next Acts of Terrorism," *USA Today,* September 24, 2001, p. A1.

47. Sheryl Gay Stolberg, "A Nation Challenged: The Biological Threat," *New York Times,* September 30, 2001.

48. Eisler, Bayles, and Vergano, "U.S. Cities Brace for the Next Act of Terrorism."

49. Associated Press, "White House Faces Disclosure Suit," *Washington Post,* June 8, 2002, p. A11.

50. Tamar Lewin, "A Nation Challenged: Fear of Infections," *New York Times,* September 27, 2001.

51. Leonard A. Cole, *The Anthrax Letters: A Medical Detective Story* (Washington, DC: National Academy of Sciences, 2003), p. 50.

52. Stolberg, "A Nation Challenged: The Biological Threat."

53. "Vulnerable Home Front," *USA Today,* September 13, 2001, p. A12.

54. Melody Petersen and Andrew Pollack, "A Nation Challenged: The Defenses," *New York Times,* September 28, 2001.

55. Cole, *Anthrax Letters,* p. 116f.

56. Alessandra Stanley, "A Nation Challenged: The Discourse," *New York Times,* September 22, 2001.

57. Massimo Calabresi and Sally Donnelly, "Cropduster Manual Discovered," *Time Online Edition,* September 22, 2001, http://www.time.com/time/nation/article/0,8599,175951,00.html.

58. These rumors are discussed extensively in Marylin W. Thompson, *The Killer Strain: Anthrax and a Government Exposed* (New York: HarperCollins, 2003).

59. Denny Lee, "Shadows across the City," *New York Times,* September 23, 2001.

60. Maureen Dowd, "Liberties, from Botox to Botulism," *New York Times,* September 26, 2001.

61. Ibid.

62. Lewin, "A Nation Challenged: Fear of Infections."

63. Rita Rubin, "Pediatricians Getting Queries on Smallpox, Anthrax," *USA Today,* October 3, 2001, p. D8.

64. Greg Winter and William J. Broad, "The Water Supply: Added Security for Dams, Reservoirs and Aqueducts," *New York Times,* September 26, 2001.

65. Petersen and Pollack, "A Nation Challenged: The Defenses."

66. Paul Leavitt, "New Postage Stamp Offers Unity Message," *USA Today,* October 3, 2001, p. A7.

67. "Anthrax Killer 'Is US Defence Insider,'" *BBC News Online,* August 18, 2002, http://news.bbc.co.uk/1/hi/world/americas/2196008.stm.

68. "Das Werk eines Profis," *Süddeutsche Zeitung,* November 28, 2003, p. 13.

69. Robert Graysmith, *Amerithrax: The Hunt for the Anthrax Killer* (New York: Berkley, 2004), pp. 444–448.

70. Philip Shenon, "9/11 Commission Says U.S. Agencies Slow Its Inquiry," *New York Times,* July 9, 2003.

71. "Wrestling for the Truth of 9/11," *New York Times,* July 9, 2003, p. A20.

72. Kean, *9/11 Commission Report,* p. 151.

73. Ibid., p. 490. JI is "Jemaah Islamiya, a nascent organization headed by Indonesian Islamists with cells scattered across Malaysia, Singapore, Indonesia, and the Philippines" (p. 58).

74. Martin Schütz, email message to author, July 17, 2003.

5. Foreign Bodies

1. *Anthrax,* "News," http://anthrax.com/nfws.htm (accessed January, 2002).

2. Ibid. (accessed February 2003).

3. František Graus, *Pest—Geissler—Judenmorde: Das 14. Jahrhundert als Krisenzeit* (Göttingen: Vandenhoeck and Ruprecht, 1987), p. 302.

4. Barbara Dettke, *Die asiatische Hydra: Die Cholera von 1830/31 in Berlin und den preußischen Provinzen Posen, Preußen und Schlesien* (Berlin: de Gruyter, 1995), pp. 294–295.

5. Ibid. p. 296.

6. Stefan Winkle, *Geißeln der Menschheit. Kulturgeschichte der Seuchen* (Dusseldorf: Artemis and Winkler, 1997), pp. 159–187.

7. See the classic paper by Erwin H. Ackerknecht, "Anticontagionism between 1821 and 1867," *Bulletin of the History of Medicine* 22 (1948), pp. 562–593; see also Margaret Pelling, "Contagion/Germ Theory/Specificity," in W. F. Bynum and Roy Porter,

Companion Encyclopedia of the History of Medicine (London: Routledge, 1993), vol. 1, pp. 309–334.

8. Detailed in Olaf Briese, *Angst in den Zeiten der Cholera,* vol. 1, *Über kulturelle Ursprünge des Bakteriums* (Berlin: Akademie, 2003), pp. 211ff.

9. Cited in Dettke, *Die asiastische Hydra,* p. 271.

10. August Theodor Stamm, "Einige Bemerkungen über die Entstehungsursachen und über die Vernichtungsmöglichkeiten epidemischer Krankheiten," in *Versammlung deutscher Naturforscher und Ärzte* 37 (Berlin: n.p., 1862): 83–85.

11. All citations from, respectively, *Deutsche Medicinische Wochenschrift,* no. 31 (1869): 330; no. 19 (1879): 241; no. 39 (1879): 499–500; no. 50 (1879): 642; no. 22 (1880): 291; *Berner Tagblatt,* September 28, 1908.

12. Tom Clancy, *Rainbow Six* (New York: Berkley, 1998), p. 125.

13. See also Bruno Latour, *The Pasteurization of France* (Cambridge: Harvard University Press, 1988).

14. Ferdinand Cohn, *Ueber Bakterien, die kleinsten lebenden Wesen* (Berlin: Lüderitz'sche, 1872), p. 12.

15. Robert Koch, *Untersuchungen über die Aetiologie der Wundinfectionskrankheiten* (Leipzig: Vogel, 1878), p. 79.

16. Elie Metchnikoff, "Sur la lutte des cellules de l'organisme contre l'invasion des microbes (théorie des phagocytes)," *Annales de l'Institut Pasteur* 1 (1887), pp. 323, 324, 328; Alfred I. Tauber and Leon Chernyak, *Metchnikoff and the Origins of Immunology: From Metaphor to Theory,* Monographs on the History and Philosophy of Biology (New York: Oxford University Press, 1991).

17. Arthur M. Silverstein, "Cellular versus Humoral Immunity:

Determinants and Consequences of an Epic Nineteenth Century Battle," *Cellular Immunology* 48 (1979): 208–221.

18. Emil Behring, "Untersuchungen über das Zustandekommen der Diphtherie-Immunität bei Thieren," *Deutsche Medicinische Wochenschrift* 50 (1890): 1145–1148.

19. Christoph Gradmann, "Invisible Enemies: Bacteriology and the Language of Politics in Imperial Germany," *Science in Context* 13 (2000): 9–30; ibid., "Robert Koch and the Pressures of Scientific Research: Tuberculosis and Tuberculin," *Medical History* 45 (2001): 1–32; Laura Otis, *Membranes: Metaphors of Invasion in Nineteenth-Century Literature, Science, and Politics* (Baltimore: Johns Hopkins University Press, 1999); Roy Porter, *The Greatest Benefit to Mankind: A Medical History of Mankind* (London: Harper Collins, 1997), pp. 429–461.

20. Edward Albert Schäfer, *Das Leben: Sein Wesen, sein Ursprung und seine Erhaltung* (Berlin: Julius Springer, 1913), p. 60.

21. *Le Journal médical quotidien*, 61 (1885): 3, cited in Rudolf Virchow, "Der Kampf der Zellen und der Bakterien," *Archiv für pathologische Anatomie und Physiologie und für klinische Medizin* 101 (1885): 1–13.

22. Peter M. Holt, *The Mahdist State in the Sudan, 1881–1898*, 2nd ed. (Oxford: Clarendon, 1970); Michael Barthorp, *Blood-Red Desert Sand: The British Invasions of Egypt and the Sudan 1882–98* (London: Weidenfeld and Nicholson, 2002).

23. See Sven Lindqvist, *Exterminate All the Brutes* (London: Granta Books, 2002).

24. Paul Weindling, *Epidemics and Genocide in Eastern Europe, 1890–1945* (Oxford: Oxford University Press, 2000), pp. 58–62.

25. Paul Weindling, "A Virulent Strain: German Bacteriology as Scientific Racism, 1890–1920," in *Race, Science, and Medicine, 1700–1960,* ed. B. Harris and E. Waltraud (London: Routledge, 1999), pp. 218–234; cf. Sarah Jansen, *"Schädlinge": Geschichte eines wissenschaftlichen und politischen Konstrukts, 1840–1920* (Frankfurt: Campus, 2003), pp. 249–255.

26. My thanks for this reference to Silvia Berger, who has been exploring German bacteriology in the First World War and trends in germ paradigms from 1900 to1930 in the context of our research project, "Political Metaphors in Bacteriology and Immunology, 1870–1930."

27. J. Andrew Mendelsohn, "Von der 'Ausrottung' zum Gleichgewicht: Wie Epidemien nach dem Ersten Weltkrieg komplex wurden," in *Strategien der Kausalität: Konzepte der Krankheitsverursachung im 19. und 20. Jahrhundert,* ed. Christoph Gradmann and Thomas Schlich (Pfaffenweiler: Cantaurus, 1999), pp. 226–268, quotation p. 246.

28. Ministry of Health (Great Britain), *Report on the Pandemic of Influenza, 1918–19* (London, 1920), cited in Mendelsohn, "Von der 'Ausrottung' zum Gleichgewicht," p. 247.

29. Ludwik Fleck, *Entstehung und Entwicklung einer wissenschaftlichen Tatsache: Einführung in die Lehre vom Denkstil und Denkkollektiv* (Basel: Schwabe, 1935; reprint, Frankfurt: Suhrkamp, 1993), p. 79. For a similar reference to the term invasion, see Owsei Temkin, "An Historical Analysis of the Concept of Infection," in *The Double Face of Janus and Other Essays in the History of Medicine* (Baltimore: Johns Hopkins University Press, 1977), pp. 456–471, esp. p. 456.

30. Francisco J. Varela, "Der Körper denkt: Das Immunsystem und der Prozess der Körper-Individuierung," in *Paradoxien, Dissonanzen, Zusammenbrüche: Situationen offener Epistemologie,* ed. Hans Ulrich Gumbrecht and K. Ludwig Peiffer (Frankfurt: Suhrkamp, 1991), pp. 727–743.

31. Joshua Lederberg, "Infectious History," *Science* 288, no. 5464 (April 14, 2000): 287–293.

32. Jansen, *"Schädlinge,"* p. 337.

33. Alex Bein, "The Jewish Parasite: Notes on the Semantics of the Jewish Problem with Special Reference to Germany," *Leo Baeck Institut, Year Book* 9 (1964): 3–40, esp. pp. 34–35.

34. For a more detailed version see Weindling, *Epidemics and Genocide,* pp. 292–315.

35. On the history of the term *parasite,* see Ulrich Enzensberger, *Parasiten: Ein Sachbuch* (Frankfurt: Eichborn, 2001).

36. Hans-Lukas Kieser and Dominik Schaller, eds., *Der Völkermord an den Armeniern und die Shoa* [*The Armenian Genocide and the Shoa*] (Zurich: Chronos, 2002), pp. 11–80.

37. Cited in Gerd Koenen, *Utopie der Säuberung: Was war der Kommunismus?* (Frankfurt: Fischer, 2000), p. 63.

38. Cited in ibid., p. 64f.

39. Ibid.

40. Edmund P. Russell, III: "'Speaking of Annihilation': Mobilizing for War against Human and Insect Enemies 1914–1945," *Journal of American History* 82, no. 4 (March 1996): 1505–1529. See Norman M. Naimark, *Fires of Hatred: Ethnic Cleansing in Twentieth-Century Europe* (Cambridge: Harvard University Press, 2001); Barrington Moore, *Moral Purity and Persecution in History* (Princeton: Princeton University Press, 2000).

6. Infection, the Metaphor of Globalization

1. Rainer Prätorius, *In God We Trust: Religion und Politik in den USA* (Munich: Beck, 2003).

2. Nancy Tomes, *The Gospel of the Germ: Men, Women, and the Microbe in American Life* (Cambridge: Harvard University Press, 1998).

3. Ellen Schrecker, *Many Are the Crimes: McCarthyism in America* (Princeton: Princeton University Press, 1998), p. 144; Eric Bentley, *Thirty Years of Treason: Excerpts from Hearings before the House Committee on Un-American Activities, 1938–1968* (New York: Thunder's Mouth Press, 1971), p. 147; J. Edgar Hoover, *On Communism* (New York: Random House, 1969), p. 128 [cited in Thomas Mergel, "The Unknown and the Familiar Enemy: The Semantics of Anticommunism in the U.S. and Germany, 1945–1970," in *Languages of Propaganda,* ed. Willibald Steinmetz (Oxford: Oxford University Press, 2006)].

4. Francis Fukuyama, "The End of History," *National Interest* 16 (1989): 3–18; Emmanuel Todd, *After the Empire: The Breakdown of the American Order* (New York: Columbia University Press, 2004).

5. Interview with a man on the street immediately after the September 11 terrorist attack, New York, September 11, 2001, cited in Sandra Silberstein, *War of Words: Language, Politics, and 9/11* (London: Routledge, 2002), p. 66.

6. See http://911.swafford-family.com/911-34b.html.

7. See http://www.toynutz.com/AwakenedEagle.html.

8. See http://home.comcast.net/~bill.fisher/911attack.html.

9. Deroy Murdock, "Trampling Terrorists: A How-To Guide,"

National Review Online, October 19, 2001, http://www.nationalreview.com/murdock/murdock101901.shtml.

10. Gary Aldrich, "Take Terrorist's Rights, Not Ours!" *WorldNetDaily.com,* September 28, 2001, http://www.worldnetdaily.com/news/article.asp?ARTICLE_ID=24714.

11. UN General Assembly, 57th Session, in Plenary Session, "Statement by Ambassador Richard S. Williamson, United States Alternate Representative to the United Nations, on Afghanistan," press release #212 (02), December 6, 2002.

12. Alan Tice, "Lessons from the Taliban for Infection Control," *Infectious Disease News,* December 2001, www.infectious-diseasenews.com/200112/guested.asp.

13. President George W. Bush, "Address before a Joint Session of the Congress on the United States Response to the Terrorist Attacks of September 11, September 20, 2001," *Weekly Compilation of Presidential Documents* 37, no. 38 (September 24, 2001): 1349.

14. Ibid., pp. 1347–1350.

15. Ibid.

16. Prime Minister of the United Kingdom Tony Blair, "Address to the U.S. Congress," July 17, 2003, 109th Congress, 1st session, *Congressional Record* 149 (2003): H7060.

17. Paul Stares and Mona Yacoubian, "Terrorism as Virus," *Washington Post,* August 23, 2005, p. A15.

18. U. S. Senate Committee on the Judiciary, *Homeland Defense: Hearing before the Committee on the Judiciary,* 107th Congress, 1st session, September 25, 2001.

19. For the concept of "civilization" as a common racist phantasm, see Gazi Çağlar, *Der Mythos vom Krieg der Zivilisationen:*

Der Westen gegen den Rest der Welt. Eine Replik auf Samuel P. Huntingtons Kampf der Kulturen (Münster: Unrast-Verlag, 2002).

20. Michel Foucault, *Society Must Be Defended: Lectures at the College de France (1975–76),* trans. David Macey (New York: Picador, 2003), p. 255.

21. Ibid., p. 296.

22. See Bush, "Address," *Weekly Compilation of Presidential Documents* 37, no. 38 (September 24, 2001): 1347–1351.

23. Cited in Sven Lindqvist, *Exterminate All the Brutes: A Modern Odyssey into the Heart of Darkness,* trans. Joan Tate (New York: New Press, 1997), p. 145.

24. Carl Schmitt (1922): *Politische Theologie: Vier Kapitel zur Lehre der Souveränität,* cited in Giorgio Agamben, *Homo sacer: Sovereign Power and Bare Life* (Stanford: Stanford University Press, 1998), p. 11.

25. See, for example, http://www.toynutz.com/AwakenedEagle.html.

26. Derrick Z. Jackson, "Noble Act or Political Assassination?" *Boston Globe,* July 25, 2003.

27. Agamben, *Homo sacer.*

28. Ibid., p. 7.

29. Tom Clancy, *Rainbow Six* (New York: Berkley, 1998), p. 871.

30. Ibid., pp. 836, 842.

31. Ibid., pp. 858f.

32. Ibid., pp. 889, 895.

33. Michel Foucault, *Power/Knowledge: Selected Interviews and Other Writings, 1972–1977,* ed. Colin Gordon (New York: Pantheon, 1972).

34. Clancy, *Rainbow Six,* p. 879.

35. *The National Security Strategy of the United States of America, September 2002* (Washington: The White House, 2002), p. 6f.

36. Samuel Huntington, *Who Are We? The Challenges to America's National Identity* (New York: Simon & Schuster, 2004).

37. Angela Gonzales and Mike Sunnucks, "Illegals Bring New Disease Outbreaks! Who's Cooking Your Food?" *Business Journal of Phoenix,* May 5, 2005.

38. Joyce Howard Price, "Disease, Unwanted Import," *Washington Times Online,* February 13, 2005, http://washingtontimes.com/specialreport/20050212-112200-6485r.htm.

39. Donald G. McNeil, Jr., "Rare Infection Threatens to Spread in Blood Supply," *New York Times,* November 18, 2003, p. 1.

40. Robert Klein Engler, "Immigration and Disease: It's Enough to Make You Sick," *American Daily* (Phoenix, AZ), November 21, 2003.

41. *Superhuman: Killers into Cures,* vol. 2, VHS, produced by Michael Mosley (London: BBC, 2000).

42. See http://www.canadafirst.net/immi-kill.

43. "Deadlier Than Any Virus: Euro-Doctors Pass Secret Anti-Immigrant Motion," September 9, 1997, http://www.survivreausida.net/article2647.html.

44. H. G. Wells, *War of the Worlds* (New York: Harper and Brothers, 1898; reprint, 2003), p. 180.

45. Ibid., p. 181.

46. Ibid., p. 5.

47. Ibid.

48. Michael Hardt and Antonio Negri, *American Empire* (Cambridge: Harvard University Press, 2000).

49. Michel Foucault, *Discipline and Punish: The Birth of the Prison* (London: Penguin, 1991), p. 195.

50. Ibid., pp. 197, 198.

51. Ibid.

52. Ibid., p. 200f.

53. Jack Z. Bratich, Jeremy Packer, and Cameron McCarthy, eds., *Foucault, Cultural Studies, and Governmentality* (Albany: State University of New York Press, 2003).

54. *United for a Stronger America: Citizens' Preparedness Guide* (Washington: National Crime Prevention Council, 2002), p. 2.

55. Ibid., p. 18.

56. Francine Prose, "Die Schule der Angst," *Die Zeit,* no. 32 (2003).

57. Nils Leopold, "Aufgeklärte Politik öffentlicher Sicherheit oder symbolischer Krieg gegen das Böse? Eine Analyse der Anti-Terror-Gesetzgebung," *Vorgänge: Zeitschrift für Bürgerrechte und Gesellschaftspolitik* 41 (2002), pp. 31–40; Katharina Sophie Rürup, "Bürgerrechte ade? Die Gesetzgebung in den USA nach dem 11. September," *Vorgänge: Zeitschrift für Bürgerrechte und Gesellschaftspolitik* 41 (2002), pp. 52–60.

58. Marouf A. Hasian, "Power, Medical Knowledge, and the Rhetorical Invention of 'Typhoid Mary,'" *Journal of Medical Humanities* 21 (2000): 123–139; Priscilla Wald, "'Typhoid Mary' and the Science of Social Control," *Social Text* 15 (1997): 181–214.

59. Reference courtesy of Dave Schläpfer, who is working on a dissertation on the persistence of Tyhpoid Mary in popular culture.

60. The band is currently known as The Revelevens; I would like to thank Carrie Donovan for permission to quote her text.

61. Blood, "Ebola," *Gas, Flames, Bones,* CD (1999).

62. Thought Riot, "Sepsis, Part 1," *The Dangerous Doctrine of Empathy,* CD (2003).

63. Thee Maldoror Kollective, interview by strennus, 2002, http://strenuus.interfree.it/interviews/2002_theemaldororkollective.htm.

64. Alice Cooper, "Nuclear Infected," *Flush the Fashion,* CD (Burbank: Warner Bros.: 1997).

65. See http://www.guerrillagirls.com/posters/venicewalld.shtml.

66. See http://www.cockeyed.com/citizen/alert/action.html.

67. See the game maker's official website for a description of this game: http://www.dothack.com/game/index.html. See the Cannes Film Festival's website for a description of the movie: http://www.festival-cannes.fr.

68. *28 Days Later,* DVD, directed by Danny Boyle (2002; Los Angeles: Twentieth Century Fox, 2003).

69. The next several quotes are all from *Twelve Monkeys,* DVD, directed by Terry Gilliam (1995; Universal City: MCA Home Video, 1998).

70. See Slajov Žižek, *Welcome to the Desert of the Real: Five Essays on September 11 and Related Dates* (London: Verso, 2002).

71. Phillip Matanka, "Sales of Hoax Anthrax on the Rise," *Online Newspaper Gazette,* December 31, 2001, http://craptaculus.com/News/business/fake_anthrax.shtml.

Epilogue: Smallpox Liberalism

1. Rob Stein, "Pox-Like Outbreak Reported," *Washington Post,* June 8, 2003, p. A1.

2. *Neue Zürcher Zeitung,* June 10, 2003, p. 14.

3. "An Open Letter to Elias Zerhouni," *Science* 307, no. 5714 (March 2005), pp. 1409–1410.

4. "Wir sollten vorbereitet sein," interview with Ken Alibek, *NZZ am Sonntag,* June 5, 2005.

5. Andreas Bubnoff, "The 1918 Flu Virus Is Resurrected," *Nature* 437 (October 6, 2005): 794–795; Terrence M. Tumpey, et al., "Characterization of the Reconstructed 1918 Spanish Influenza Pandemic Virus," *Science* 310, no. 5745 (October 7, 2005): 77–80.

6. Raphael Perl, *Terrorism and National Security: Issues and Trends,* CRS Report for Congress, June 8, 2005.

7. Michel Foucault, *Politics, Philosophy, Cultures: Interviews and Other Writings, 1977–1984,* ed. Laurence D. Kritzman, trans. Alan Sheridan (London: Routledge, Chapman and Hall, 1988), p. 209.

8. Michel Foucault, *Geschichte der Gouvernementalität I: Sicherheit, Territorium, Bevölkerung. Vorlesung am Collège de France 1977–1978,* ed. Michel Sennelart (Frankfurt: Suhrkamp, 2004), p. 79.

9. Foucault, *Geschichte der Gouvernementalität I,* p. 25.

10. Michel Foucault, *Dits et écrits 1954–1988,* vol. 3, ed. Daniel Defert and François Ewald (Paris: Gallimard, 1994), p. 467.

11. Foucault, *Geschichte der Gouvernementalität I,* p. 25.

12. Joshua Lederberg, "Infectious History," *Science* 288, no. 5464 (April 14, 2000): 287–293.

13. Paul Stares and Mona Yacoubian, "Terrorism as Virus," *Washington Post,* August 23, 2005, p. A15.

14. Robert Koch, "The Etiology of Anthrax, Based on the Life

History of *Bacillus anthracis,*" in *Milestones in Microbiology: 1556 to 1940,* trans. and ed. Thomas D. Brock (Washington, DC: ASM Press, 1998), p. 90.

15. Andrew Mendelsohn, "From Eradication to Equilibrium: How Epidemics Became Complex after World War I," in *Greater than the Parts: Holism in Biomedicine, 1920–1950,* ed. Christopher Lawrence and George Weisz (Oxford: Oxford University Press, 1998), pp. 303–331; Warwick Anderson, "Natural History of Infectious Disease: Ecological Vision in Twentieth-Century Biomedical Science," *OSIRIS* 19 (2004): 39–61.

16. Stares and Yacoubian, "Terrorism as Virus."

17. Ibid.

Acknowledgments

My thanks to everyone who made this book possible. For the updated American edition, I had the benefit of Thomas Langer's in-depth research. I am deeply indebted to Daniele Ganser, senior researcher at the Center for Security Studies at the Swiss Federal Institute of Technology (ETH), Zurich, for his insights into the events leading up to September 11 and the analysis of the Kean Report. He was most generous in sharing his expertise with me. Myriam Dunn and Reto Wollenmann brought me up to date on the bioterror threat; together with Professor Andreas Wenger they organized an international workshop on bioterror in April 2005 at the Center for Security Studies at the ETH. I would also like to thank Thomas Mergel for making available to me his text on political language in the United States and in Germany during the Cold War, Luca Canellotto for his help on the culture of computer games, and David Schläpfer for his tips on underground music. Elisabeth Bronfen, David Gugerli, Nicole Schwager, Silvia Berger, Marianne Hänseler, Myriam Spörri, and Samir Jamal Aldin all provided most helpful suggestions. Giselle Weiss cleverly managed to translate my tricky German text into a wonderfully readable American one.

Finally, this book would not have come about without the help of Bruno Latour and Michael Hagner, not to mention the much-appreciated encouragement of my editor, Michael Fisher, and the editorial assistance of Ian Stevenson. My deepest thanks go to Regula Bochsler for sharing her knowledge and also for urging me to write this essay instead of sitting on the couch gritting my teeth over CNN and global politics. She was most gracious in ignoring the aura of "anthrax" panic that sometimes reigned over our household.

Index